6:00
June 25
either Eatery
or
Pocket Rest.

I MAG INE

How Creativity Works

Jonah Lehrer

HOUGHTON MIFFLIN HARCOURT
BOSTON • NEW YORK

For information about permission to reproduce
selections from this book, write to Permissions,
Houghton Mifflin Harcourt Publishing Company,
215 Park Avenue South, New York, New York 10003.

www.hmhbooks.com

Library of Congress Cataloging-in-Publication Data
Lehrer, Jonah.
Imagine: how creativity works / Jonah Lehrer.
p. cm.
ISBN 978-0-547-38607-2
1. Creative ability. 2. Creative thinking. 3. Imagination. I. Title.
BF408.L455 2012
153.3'5 — dc23
2011044597

Book design by Alex Camlin
Illustrations by Bruce Mau Design

Printed in the United States of America
DOC 10 9 8 7 6 5 4 3 2

Excerpt from "September 1, 1939" © 1939 by W. H. Auden, renewed.
Reprinted by permission of Curtis Brown, Ltd.

For Sarah and Rose

CONTENTS

Hell is a place where nothing connects with nothing.

—T. S. Eliot, Introduction to Dante's *Inferno*

INTRODUCTION

Procter and Gamble had a problem: it needed a new floor cleaner. In the 1980s, the company had pioneered one lucrative consumer product after another, from pull-up diapers to anti-dandruff shampoo. It had developed color-safe detergent and designed a quilted paper towel that could absorb 85 percent more liquid than other paper towels. These innovations weren't lucky accidents: Procter and Gamble was deeply invested in research and development. At the time, the corporation had more scientists on staff than any other company in the world, more PhDs than the faculties of MIT, UC-Berkeley, and Harvard *combined*.

And yet, despite the best efforts of the chemists in the household-cleaning division, there were no new floor products in the pipeline. The company was still selling the same lemon-scented detergents and cloth mops; consumers were still sweeping up their kitchens using wooden brooms and metal dustpans. The reason for this creative failure was simple: it was extremely difficult to make a stronger floor cleaner that didn't also damage the floor. Although Procter and Gamble had invested millions of dollars in a new generation of soaps, these products tended to fail during the rigorous testing phase, as they peeled off wood varnishes and irritated delicate skin. The chemists assumed that they had exhausted the chemical possibilities.

That's when Procter and Gamble decided to try a new approach. The company outsourced its innovation needs to Continuum, a design firm with offices in Boston and Los Angeles. "I think P and G came to us because their scientists were telling them to give up," says Harry West, a leader on the soap team and now Continuum's CEO. "So they told us to think crazy, to try to come up with something that all those chemists couldn't."

But the Continuum designers didn't begin with molecules. They didn't spend time in the lab worrying about the chemistry of soap. Instead, they visited people's homes and watched dozens of them engage in the tedious ritual of floor cleaning. The designers took detailed notes on the vacuuming of carpets and the sweeping of kitchens. When the notes weren't enough, they set up video cameras in living rooms. "This is about the most boring footage you can imagine," West says. "It's movies of mopping, for God's sake. And we had to watch hundreds of hours of it." The videotapes may have been tedious, but they were also essential, since West and his team were trying to observe the act of floor cleaning without any preconceptions. "I wanted to forget everything I knew about mops and soaps and brooms," he says. "I wanted to look at the problem as if I'd just stepped off a spaceship from Mars."

After several months of observation—West refers to this as the anthropologist phase—the team members had their first insight. It came as they watched a woman clean her mop in the bathtub. "You've got this unwieldy pole," West says. "And you are splashing around this filthy water trying to get the dirt out of a mop head that's been expressly designed to *attract* dirt. It's an extraordinarily unpleasant activity." In fact, when the Continuum team analyzed the videotapes, they found that people spent more time cleaning their mops than they did cleaning the floors; the tool made the task more difficult. "Once I realized how bad mopping was, I became quite passionate about floor cleaning," West says. "I became convinced that the world didn't need an improved version of the mop. Instead, it needed a total *replacement* for the mop. It's a hopeless piece of technology."

Unfortunately, the Continuum designers couldn't think of a better cleaning method. It seemed like an impossible challenge. Perhaps floor cleaning was destined to be an inefficient chore.

In desperation, the team returned to making house visits, hoping for some errant inspiration. One day, the designers were watching an elderly woman sweep some coffee grounds off the kitchen floor. She got out her hand broom and carefully brushed the grounds into a dustpan. But then something interesting happened. After the woman was done sweeping, she wet a paper towel and wiped it over the linoleum, picking up the last bits of spilled coffee. Although everyone on the Continuum team had done the same thing countless times before, this particular piece of dirty paper led to a revelation.

What the designers saw in that paper towel was the possibility of a disposable cleaning surface. "All of a sudden, we realized what needed to be done," says Don Buchner, a Continuum vice president. "We needed to invent a spot cleaner that people could just throw away. No more cleaning mop heads, no more bending over in the bathtub, no more buckets of dirty water. *That* was our big idea." A few weeks later, this epiphany gave rise to their first floor-cleaning prototype. It was a simple thing, just a slender plastic stick connected to a flat rectangle of Velcro to which disposable pieces of electrostatic tissue were attached. A spray mechanism was built into the device, allowing people to wet the floor with a mild soap before they applied the wipes. (The soap was mostly unnecessary, but it smelled nice.) "You know an idea has promise when it seems obvious in retrospect," West says. "Why splash around dirty water when you can just wipe up the dirt? And why would you bother to clean this surface? Why not just throw it away, like a used paper towel?"

Procter and Gamble, however, wasn't thrilled with the concept. The company had developed a billion-dollar market selling consumers the latest mops and soaps. They didn't want to replace that business with an untested cleaning product. The first focus groups only reinforced the skepticism. When Procter and Gamble presented consumers with a sketch of the new cleaning device, the vast majority of people rejected the concept. They didn't want to throw out their mops or have to rely on a tool that was little more than a tissue on a stick. They didn't like the idea of disposable wipes, and they didn't understand how all that dirt would get onto the moistened piece of paper. And so the idea was shelved; Procter and Gamble wasn't going to risk market share on a radical new device that nobody wanted.

But the designers at Continuum refused to give up—they were convinced they'd discovered the mop of the future. After a year of pleading, they persuaded Procter and Gamble to let them show their prototype to a focus group. Instead of just reading a description of the product, consumers could now play with an "experiential model" clad in roughly cut plastic. The prototype made all the difference: people were now enthralled by the cleaning tool, which they tested out on actual floors. In fact, the product scored higher in focus-group sessions than any other cleaning device Procter and Gamble had ever tested. "It was off the charts," Buchner says. "The same people who hated the idea when it was just an idea now wanted to take the thing home with them." Furthermore, tests by Procter and Gamble demonstrated that the new product cleaned the floor far better than sponge mops, string mops, or any other kinds of mops. According to the corporate scientists, the "tissue on a stick" was one of the most effective floor cleaners ever invented.

In 1997, nearly three years after West and his designers began making their videotapes, Procter and Gamble officially submitted an application for a U.S. patent. In the early spring of 1999, the new cleaning tool was introduced in supermarkets across the country. The product was an instant success: by the end of the year, it had generated more than $500 million in sales. Numerous imitators and spinoffs have since been introduced, but the original device continues to dominate the post-mop market, taking up an ever greater share of the supermarket aisle. Its name is the Swiffer.

The invention of the Swiffer is a tale of creativity. It's the story of a few engineers coming up with an entirely new cleaning tool while watching someone sweep up some coffee grounds. In that flash of thought, Harry West and his team managed to think differently about something we all do every day. They were able to see the world as it was—a frustrating place filled with tedious chores—and then envision the world as it might be if only there were a better mop. That insight changed floor cleaning forever.

This book is about how such moments happen. It is about our most important mental talent: the ability to imagine what has never existed. We take this talent for granted, but our lives are defined by it. There is the pop song on the radio and the gadget in your pocket, the art on the wall and the air conditioner in the window. There is the medicine in the bathroom and the chair you are sitting in and this book in your hand.

And yet, although we are always surrounded by our creations, there is something profoundly mysterious about the creative process. For instance, why did Harry West come up with the Swiffer concept after watching that woman wipe the floor with the paper towel? After all, he'd done it himself on numerous occasions. "I can't begin to explain why the idea arrived then," he says. "I was too grateful to ask too many questions." The sheer secrecy of creativity—the difficulty in understanding how it happens, even when it happens to us—means that we often associate breakthroughs with an external force. In fact, until the Enlightenment, the imagination was entirely synonymous with higher powers: being creative meant channeling the muses, giving voice to the ingenious gods. (*Inspiration,* after all, literally means "breathed upon.") Because people couldn't understand creativity, they assumed that their best ideas came from somewhere else. The imagination was outsourced.

The deep mysteriousness of creativity also intimidated scientists. It's one thing to study nerve-reaction times or the mechanics of sight. But how does one measure the imagination? The daunting nature of the subject led researchers to mostly neglect it; a recent survey of psychology papers published between 1950 and 2000 revealed that less than 1 percent of them investigated aspects of the creative process. Even the evolution of this human talent was confounding. Most cognitive skills have elaborate biological histories, so their evolution can be traced over time. But not creativity—the human imagination has no clear precursors. There is no ingenuity module that got enlarged in the human cortex, or even a proto-creative impulse evident in other primates. Monkeys don't paint; chimps don't write poems; and it's the rare animal (like the New Caledonian crow) that exhibits rudimentary signs of problem solving. The birth of creativity, in other words, arrived like any insight: out of nowhere.

This doesn't mean, however, that the imagination can't be rigorously studied. Until we understand the set of mental events that give rise to new thoughts, we will never understand what makes us so special. That's why this book begins by returning us to the material source of the imagination: the three pounds of flesh inside the skull. William James described the creative process as a "seething cauldron of ideas, where everything is fizzling and bobbing about in a state of bewildering activity." For the first time, we can see the cauldron itself, that massive network of electrical cells that allow individuals to form new connections between old ideas. We can take snapshots of thoughts in brain scanners and measure the excitement of neurons as they get closer to a solution. The imagination can seem like a magic trick of matter—new ideas emerging from thin air—but we are beginning to understand how the trick works.

The first thing this new perspective makes clear is that the standard definition of *creativity* is completely wrong. Ever since the ancient Greeks, people have assumed that the imagination is separate from other kinds of cognition. But the latest science suggests that this assumption is false. Instead, *creativity* is a catchall term for a variety of distinct thought processes. (The brain is the ultimate category buster.) Just consider the profusion of creative methods that led to the invention of the Swiffer. First, there was the anthropologist phase, those nine months of careful observation and tedious videotaping. Although this phase didn't generate any new ideas—the point was to clear the mind of old ones—it played an essential role in the creative process, allowing the team to better understand the problem. And then, when West watched the woman sweep up the coffee grounds, there was the classic moment of insight, a breakthrough appearing in a fraction of a second. But that epiphany wasn't the end of the process. The engineers and designers still had to spend years fine-tuning the design, perfecting the spray nozzle and the electrostatic wipes. "The concept is only the start of the process," West says. "The hardest work always comes after, when you're trying to make the idea real."

The point is that the Swiffer creative process involved multiple forms of creativity. This is where the tools of modern science prove essential, since they allow us to see how these various forms depend on different kinds of brain activity. The imagination is transformed from something metaphysical—a property of the gods—into a particular twitch of cortex. Furthermore, this new knowledge is useful: because we finally understand what creativity is, we can begin to construct a taxonomy of it, outlining the conditions under which each particular mental strategy is ideal. Some acts of imagination are best done in a crowded café sipping espresso, and some are helped by a cold beer on the couch. Sometimes we need to let go and improvise on our own, and sometimes we need the wisdom of others. Once we know how creativity works, we can make it work for us.

But just because we've begun to decipher the anatomy of the imagination doesn't mean we've unlocked its secret. In fact, this is what makes the subject of creativity so interesting: it requires a description from multiple perspectives. The individual brain, after all, is always situated in a context and a culture, so we need to blend psychology and sociology, merging together the outside world and the inside of the mind. This is why, although *Imagine* begins with the fluttering of neurons, it will also explore the influence of the surrounding environment on creativity. Why are some cities such centers of innovation? What kind of classroom techniques increase the creativity of children? Is the Internet making us more or less imaginative? We'll look at evidence showing that seemingly irrelevant factors—such as the color of paint on the wall or the location of a restroom—can have a dramatic impact on creative production.

Furthermore, because the act of invention is often a collaborative process—we are inspired by other people—it's essential that we learn to collaborate in the right way. The first half of this book focuses on individual creativity, while the second half shows what happens when people come together, interacting in office hallways and city streets. Thanks to some fascinating new research, such as an analysis of the partnerships behind thousands of Broadway musicals, we can begin to understand why some teams and companies are so much more creative than others. Their success is not an accident.

For most of human history, people have believed that the imagination is inherently inscrutable, an impenetrable biological gift. As a result, we cling to a series of false myths about what creativity is and where it comes from. These myths don't just mislead—they also interfere with the imagination. In addition to looking at elegant experiments and scientific studies, we'll examine creativity as it is experienced in the real world. We'll learn about Bob Dylan's writing method and the drug habits of poets. We'll spend time with a bartender who thinks like a chemist, and an autistic surfer who invented a new surfing move. We'll look at a website that helps solve seemingly impossible problems, and we'll go behind the scenes at Pixar. We'll watch Yo-Yo Ma improvise, and we'll uncover the secrets of consistently innovative companies.

The point is to collapse the layers of description separating the nerve cell from the finished symphony, the cortical circuit from the successful product. Creativity shouldn't be seen as something otherworldly. It shouldn't be thought of as a process reserved for artists and inventors and other "creative types." The human mind, after all, has the creative impulse built into its operating system, hard-wired into its most essential programming code. At any given moment, the brain is automatically forming new associations, continually connecting an everyday x to an unexpected y. This book is about how that happens. It is the story of how we imagine.

1 BOB DYLAN'S BRAIN

Always carry a light bulb.
—Bob Dylan

BOB DYLAN LOOKS bored. It's May of 1965 and he's slumped in a quilted armchair at the Savoy, a fancy London hotel. His Ray-Bans are pulled down low; his eyes stuck in a distant stare. The camera turns away—Dylan's weariness feels like an accusation—and starts to pan around the room, capturing the ragged entourage of folkies and groupies following the singer on the final week of his European tour.

For the previous four months, Dylan had been struggling to maintain a grueling performance schedule. He'd traveled across the Northeast of the United States on a bus, playing in small college towns and big-city theaters. (Dylan played five venues in New Jersey alone.) Then he crossed over to the West Coast and crammed in a hectic few weeks of concerts and promotion. He'd been paraded in front of the press and asked an endless series of inane questions, from "What is the truth?" to "Why is there a cat on the cover of your last album?" At times, Dylan lost his temper and became obstinate with reporters. "I've got nothing to say about these things I write," he insisted. "I just write them. There's no great message. Stop asking me to explain." When Dylan wasn't surly, he was often sarcastic, telling journalists that he collected monkey wrenches, that he was born in Acapulco, and that his songs were inspired by "chaos, watermelons, and clocks." That last line almost made him smile.

By the time Dylan arrived in London, it was clear that the trip was taking a toll. The singer was skinny from insomnia and pills; his nails were yellow from nicotine; and his skin had a ghostly pallor. (He looked, someone said, like an "underfed angel.") Dylan was taking too many drugs and was surrounded by too many people taking drugs. In a classic scene from *Don't Look Back*, a documentary about the 1965 tour by D. A. Pennebaker, the singer returns to an empty suite. "Welcome home," says a member of his entourage. "It's the first time that this room hasn't been full of a bunch of insane lunatics, man, that I can remember . . . It's the first time it's been *cool* around here." A few minutes later, there's a knock on the door. The lunatics have arrived.

Dylan couldn't escape from the crowds, so he learned to disappear into himself. He packed a typewriter in with his luggage and could turn anything into a desk; he searched for words while surrounded by the distractions of touring. When he got particularly frustrated, he would tear his work into smaller and smaller pieces, shredding them and throwing them in the wastebasket. (Marianne Faithfull referred to such moments as "tantrums of genius.") Although Dylan's creativity remained a constant—he wrote because he didn't know what else to do—there were increasing signs that he was losing interest in creating music. For the first time, his solo shows felt formulaic, as if he were singing the lines of someone else. He rarely acknowledged the audience or paused between songs; he seemed to be in a hurry to get off-stage. In *Don't Look Back,* when a fan tells Dylan she doesn't like his new single — it featured an electric guitar — his reply is withering: "Oh, you're one of those. I understand now." And then he turns and walks away.

Before long, it all became too much. While touring in England, Dylan decided that he was leading an impossible life, that this existence couldn't be sustained. The only talent he cared about — his ceaseless creativity — was being ruined by fame. The breaking point probably came after a brief vacation in Portugal, where Dylan got a vicious case of food poisoning. The illness forced him to stay in bed for a week, giving the singer a rare chance to reflect. "I realized I was very drained," Dylan would later confess. "I was playing a lot of songs I didn't want to play. I was singing words I didn't really want to sing . . . It's very tiring having other people tell you how much they dig you if you yourself don't dig you."

In other words, Dylan was sick of his music. He was sick of strumming his acoustic guitar and standing in the spotlight by himself; sick of the politics and the expectations; sick of the burden of being a spokesman. People assumed that his songs always carried a message, that his art was really about current events. But Dylan didn't want to have an opinion on everything; he wasn't interested in being defined by the sentimental self-righteousness of "Blowin' in the Wind." The problem was that he didn't know what to do next: he felt trapped by his past but had no plan for the future. The only thing he was sure of was that this life couldn't last. Whenever Dylan read about himself in the newspaper, he made the same observation: "God, I'm glad I'm not me," he said. "I'm glad I'm not *that.*"

The last shows were in London at a sold-out Royal Albert Hall.

It was here that Dylan told his manager he was quitting the music business. He was finished with singing and songwriting and was going to move to a tiny cabin in Woodstock, New York. Although Dylan had become a pop icon—the prophetic poet of his generation—he was ready to renounce it all, to surrender the celebrity and status, if it meant he might be left alone.

Dylan wasn't bluffing. As promised, he returned from his British tour and rode his Triumph motorcycle straight out of New York City. He was leaving the folk scene of the Village behind, heading upstate to an empty house. He was done writing songs—he had nothing else to say. Dylan didn't even bring his guitar.

Every creative journey begins with a problem. It starts with a feeling of frustration, the dull ache of not being able to find the answer. We have worked hard, but we've hit the wall. We have no idea what to do next.

When we tell one another stories about creativity, we tend to leave out this phase of the creative process. We neglect to mention those days when we wanted to quit, when we believed that our problems were impossible to solve. Because such failures contradict the romantic version of events—there is nothing triumphant about a false start—we forget all about them. (The failures also remind us how close we came to having no stories to tell.) Instead, we skip straight to the breakthroughs. We tell the happy endings first.

The danger of telling this narrative is that the feeling of frustration—the act of being stumped—is an essential part of the creative process. Before we can find the answer—before we probably even know the question—we must be immersed in disappointment, convinced that a solution is beyond our reach. We need to have wrestled with the problem and lost. And so we give up and move to Woodstock because we will never create what we want to create.

It's often only at this point, *after* we've stopped searching for the answer, that the answer arrives. (The imagination has a wicked sense of irony.) And when a solution does appear, it doesn't come in dribs and drabs; the puzzle isn't solved one piece at a time. Rather, the solution is shocking in its completeness. All of a sudden, the answer to the problem that seemed so daunting becomes incredibly obvious. We curse ourselves for not seeing it sooner.

This is the clichéd moment of insight that people know so well from stories of Archimedes in the bathtub and Isaac Newton under the apple tree. It's the kind of mental process described by Coleridge and Einstein, Picasso and Mozart. When people think about creative breakthroughs, they tend to imagine them as incandescent flashes, like a light bulb going on inside the brain.

These tales of insight all share a few essential features that scientists use to define the "insight experience." The first stage is the impasse: Before there can be a breakthrough, there has to be a block. Before Bob Dylan could reinvent himself, writing the best music of his career, he needed to believe that he had nothing left to say.

If we're lucky, however, that hopelessness eventually gives way to a revelation. This is another essential feature of moments of insight: the feeling of certainty that accompanies the new idea. After Archimedes had his eureka moment—he realized that the displacement of water could be used to measure the volume of objects—he immediately leaped out of the bath and ran to tell the king about his solution. He arrived at the palace stark-naked and dripping wet.

At first glance, the moment of insight can seem like an impenetrable enigma. We are stuck and then we're not, and we have no idea what happened in between. It's as if the cortex is sharing one of its secrets.

The question, of course, is how these insights happen. What allows someone to transform a mental block into a breakthrough? And why does the answer appear when it's least expected? This is the mystery of Bob Dylan, and the only way to understand the mystery is to venture inside the brain, to break open the black box of the imagination.

1.

Mark Beeman was stumped. It was the early 1990s and Beeman, a young scientist at the National Institutes of Health, was studying patients who had suffered damage to the right hemisphere of the brain. "The doctors would always tell these people, 'Wow, you're so lucky,'" Beeman remembers. "They'd go on about how the right hemisphere was the minor hemisphere—it doesn't do much, and it doesn't do anything with language." Those consoling words reflected the scientific consensus that the right half of the brain was mostly unnecessary. In his 1981 Nobel lecture, the neuroscientist Roger Sperry summarized the prevailing view of the right hemisphere at the time he began studying it: The right hemisphere was "not only mute and agraphic but also dyslexic, word-deaf and apraxic, and lacking generally in higher cognitive function." In other words, it was thought to be a useless chunk of tissue.

But Beeman noticed that many patients with right hemisphere damage nonetheless had serious cognitive problems even though the left hemisphere had been spared. He started making a list of their deficits. The list was long. "Some of these patients couldn't understand jokes or sarcasm or metaphors," Beeman says. "Others had a tough time using a map or making sense of paintings. These might not seem like debilitating problems, but they were still very unsettling for these people, especially because they weren't supposed to exist. Their doctors had told them not to worry because the right hemisphere wasn't supposed to be important."

The struggles of these patients led Beeman to reconsider the function of the right side of the brain. At first, he couldn't figure out what all these deficits had in common. What did humor have to do with navigation? What possible link existed between sarcasm and visual art? The mental problems triggered by right hemisphere damage just seemed so incomprehensibly varied. "I couldn't come up with a decent explanation," Beeman remembers. "I couldn't connect the dots."

And then, just when Beeman was about to give up, he had an idea. Perhaps the purpose of the right hemisphere was doing the very thing he was trying to do: find the subtle connections between seemingly unrelated things.[1] Beeman realized that all of the problems experienced by his patients involved making sense of the whole, seeing not just the parts but how they hang together. "The world is so complex that the brain has to process it in two different ways at the same time," Beeman says. "It needs to see the forest *and* the trees. The right hemisphere is what helps you see the forest."

Take the language deficits caused by right hemisphere damage. Beeman speculated that, while the left hemisphere handles *denotation*—it stores the literal meanings of words—the right hemisphere deals with *connotation,* or all the meanings that can't

---

1. *This notion of the right hemisphere as a connection machine originated in the 1870s with the English neurologist John Hughlings Jackson. After studying numerous patients with injuries to the right side of the brain, Jackson concluded that, while the left hemisphere was suited for logical analysis and "willfull speech," the right hemisphere was focused on finding "associative laws."*

be looked up in the dictionary. When you read a poem or laugh at the punch line of a joke, you are relying in large part on the right hemisphere and its ability to uncover linguistic associations. Metaphors are a perfect example of this. From the perspective of the brain, a metaphor is a bridge between two ideas that, at least on the surface, are not equivalent or related. When Romeo declares that "Juliet is the sun," we know that he isn't saying his beloved is a massive, flaming ball of hydrogen. We understand that Romeo is trafficking in metaphor, calling attention to aspects of Juliet that might also apply to that bright orb in the sky. She might not be a star, but perhaps she lights up his world in the same way the sun illuminates the earth.

How does the brain understand the line "Juliet is the sun"? The left hemisphere focuses on the literal definition of the words, but that isn't particularly helpful. A metaphor, after all, can't be grasped by making a list of the adjectives that describe both entities. (In the case of *sun* and *Juliet,* that would be a very short list.) We can grasp the connection between the two nouns only by relying on their overlapping associations, by detecting the nuanced qualities they might have in common. This understanding is most likely to occur in the right hemisphere, since it's uniquely able to zoom out and parse the sentence from a more distant point of view.

This hemisphere's ability to "see the forest" doesn't apply just to language. A study conducted in the 1940s asked people with various kinds of brain damage to copy a picture of a house. Interestingly, the patients drew very different landscapes depending on which hemisphere remained intact. Patients reliant on the left hemisphere because the right hemisphere had been incapacitated depicted a house that was clearly nonsensical: front doors floated in space; roofs were upside down. However, even though these patients distorted the general form of the house, they carefully sketched its specifics and devoted lots of effort to capturing the shape of the bricks in the chimney or the wrinkles in the window curtains. (When asked to draw a person, this type of patient might draw a single hand, or two eyes, and nothing else.) In contrast, patients who were forced to rely on the right hemisphere tended to focus on the overall shape of the structure. Their pictures lacked details, but these patients got the essential architecture right. They focused on the whole.

The challenge for Beeman was finding a way to study these more abstract cognitive skills. He wanted to understand the right hemisphere—he just didn't know which questions to ask. "The right hemisphere was tainted by all this pop-psychology stuff about right-brain people being more artistic or imaginative," Beeman says. "And so when you said you wanted to investigate that kind of thinking in the right hemisphere, grant committees assumed you weren't very serious. Studying metaphors and holistic thinking seemed like a sure way to ruin a scientific career."

But in 1993, Beeman heard a talk on moments of insight by Jonathan Schooler, a psychologist now at the University of California at Santa Barbara. Schooler presented the results of a simple experiment: he'd put undergraduates in a tiny room and given them a series of difficult creative puzzles. Here's a sample question:

> A giant inverted steel pyramid is perfectly balanced on its point. Any movement of the pyramid will cause it to topple over. Underneath the pyramid is a $100 bill. How do you remove the bill without disturbing the pyramid?

Reflect, for a moment, on your own thought process as you try to solve the puzzle. Almost everyone begins by visualizing the pyramid perched precariously on the valuable piece of green paper. Your next thought probably involves some sort of crane that would lift the pyramid into the air. (Alas, such a contraption violates the rules of the puzzle.) Then you might imagine a way of sliding the money out without tearing the bill. Unfortunately, for most people, no workable solutions come to mind, which is why they reach the impasse stage. The subject gets flustered and frustrated, since he has followed his train of thought to its logical conclusion. And then he starts to give up. "One of the common reactions is for people to get annoyed at the scientist," Schooler says. "They say: 'Why'd you give me this puzzle? It's stupid. It's impossible.' You have to reassure them that the problem really has a solution."

At this point in the study, Schooler began giving the subjects hints. He subliminally flashed them a sentence with the word *fire* or told the subjects to think about the meaning of *remove*. Interestingly, these hints were much more effective when selectively presented to the left eye, which is connected to the right hemisphere. "We'd give people these funny goggles that allowed us to flash hints to one eye at a time," Schooler says. "And it was startling how you could flash a really obvious hint to the right eye [and hence left hemisphere] and it wouldn't make a difference. They still wouldn't get it. But then you'd flash the exact same hint to the other eye, and it would generate the insight. Only the right hemisphere knew what to do with the information." (If you're still wondering, the solution is to set the hundred-dollar bill on fire. The insight, then, is that the bill just needs to be *removed*, not salvaged.)

To Beeman, Schooler's finding was a revelation. It made perfect sense that the right hemisphere excelled at solving insight puzzles since that side of the brain was better able to see the hidden connections, those remote associations between separate ideas. While the left hemisphere was frantically trying to lift the pyramid into the air—that's the obvious way to "remove" the money—the right hemisphere was busy thinking about alternative approaches. "I suddenly realized that moments of insight could be a really interesting way to look at all these skills the right hemisphere excelled at," Beeman says. "It was a rigorous way to study some very mysterious aspects of the mind. I had an insight about insight."

2.

Mark Beeman has a tense smile, a receding hairline, and the wiry build of a long-distance runner. He qualified for the Olympic trials in 1988 and 1992 with a time of 3:41 in the fifteen-hundred-meter race, although he gave up competitive running after, as he puts it, "everything below the hips started to fall apart." He now subsists on long walks and the manic tapping of feet. When Beeman gets excited about something—whether it's the cellular properties of pyramidal neurons or his new treadmill—the pace of his speech accelerates and then he starts to draw pictures on whatever scratch paper is nearby.

In the mid-1990s, when Beeman began studying moments of insight, the standard scientific approach to the subject involved giving people difficult puzzles and asking them questions about how they solved them. "The problem with this method is that everything that leads you to the insight happens unconsciously," Beeman says. "People have no idea where the insight came from, or what thoughts led them to the solution. They can't tell you anything about it. The science had hit a wall."

Beeman wanted to extend the research on insight by looking at the phenomenon from the perspective of the brain. He was eager to use the new tools of modern neuroscience, such as PET scans and fMRI machines, to locate the source of epiphanies inside the skull. However, this approach immediately led to a major experimental complication. In order to isolate the brain activity that defined the insight process, Beeman needed to compare moments of insight to answers that arrived by conscious analysis, that is, by people methodically testing ideas one at a time. In conscious analysis, people have a sense of their progress and can accurately explain their thought processes. (The left hemisphere is nothing if not articulate.) The problem is solved through diligence and hard work; when the answer arrives, there is nothing sudden about it.

Unfortunately, all of the puzzles used by scientists to study insight *required* insight. Either they were solved in a sudden "Aha!" moment or they weren't solved at all. Consider this classic problem:

Marsha and Marjorie were born on the same day of the same month of the same year to the same mother and the same father, yet they are not twins. How is that possible?[2]

Or what about this one:
Rearrange the letters *n-e-w-d-o-o-r* to make one word.[3]

This was Beeman's challenge: to come up with a set of puzzles that were often solved by insight, but not always. In theory, this would allow him to isolate the unique neural patterns that defined the insight process, since he could compare the brain activity of subjects having epiphanies with that of those relying on ordinary analysis. The puzzles, though, weren't easy to invent. "It can get pretty frustrating trying to find an experimentally valid

2. *They're triplets.*
3. *The answer is "one word."*

brainteaser," Beeman says. "The puzzles can't be too hard or too easy, and you need to be able to generate lots of them." He eventually settled on a series of verbal puzzles that he named compound remote associate problems, or CRAP. The joke is beginning to get old. "Yes, yes, I'm studying CRAP," Beeman grumbles. In his science papers and PowerPoint presentations, Beeman now leaves off the final *P*.

The puzzles go like this: A subject is given three different words, such as *age, mile,* and *sand,* and asked to think of a single word that can form a compound word or phrase with each of the three. (In this case, the answer is *stone: stone age, milestone, sandstone.*) The subject has fifteen seconds to solve the question before a new puzzle appears. If he comes up with an answer, he presses the space bar on the keyboard and says whether the answer arrived via insight or analysis. When I participated in the experiment in Beeman's lab, I found that it was surprisingly easy to differentiate between these two problem-solving possibilities. When I solved puzzles with analysis, I tended to sound out each possible combination, cycling through each of the different words that went with *age* and then seeing if it also worked with *mile* and then *sand.* When I came up with a solution, I always double-checked it before pressing the space bar. An insight, by contrast, was instantaneous: the word felt like a revelation.

Beeman was now ready to start looking for the neural source of insight. He began by having people solve the puzzles while inside an fMRI machine, a brain scanner that monitors changes in blood flow as a rough correlate for changes in neural activity. (Active brain cells consume more energy and oxygen, which triggers the rush of blood.) While fMRI gives scientists a precise spatial map of the brain, the technique suffers from a time delay of several seconds while the blood diffuses across the cortex. "I soon realized that insights happen too fast for fMRI," Beeman says. "The data was just too messy."

That's when Beeman teamed up with John Kounios, a psychologist at Drexel University. Kounios's main experimental tool is EEG, or electroencephalography, which measures the waves of electricity produced by the brain. A subject wears a plastic hat filled with greased electrodes—it looks like a bulky shower cap—each of which monitors a specific frequency of neural activity. Because there is no time delay with EEG, Kounios realized that it could be a useful technique for investigating the instantaneousness of insight. Unfortunately, this speed comes at the cost of spatial resolution: the waves of electricity can't be traced back to their precise sources.

By combining both techniques—fMRI and EEG—in the same study, Beeman and Kounios were able to deconstruct the epiphany. The first thing they discovered was that, although it seemed like the answer appeared out of nowhere, the brain had been laying the groundwork for the breakthrough. (In his lectures, Beeman likes to quote a dictum of Louis Pasteur: "Chance favors the prepared mind.") The process began with an intense mental search as the left hemisphere started looking for answers in all the obvious places. Because Beeman and Kounios were giving people word puzzles, they saw additional activation in brain areas related to speech and language. This left-brain thought process, however, quickly got tiring—it took only a few seconds before the subject said he'd reached an impasse and couldn't think of the right word. "Almost all of the possibilities your brain comes up with are going to be wrong," Beeman says. "There are just so many different connections to consider. And it's up to you to keep on searching or, if necessary, change strategies and start searching somewhere else."

What happens next is the stumped phase of creativity. Not

surprisingly, this phase isn't very much fun. In the CRA study, for instance, subjects quickly got frustrated by their inability to find the necessary word. They complained to the scientists about the difficulty of the problems and threatened to quit the experiment. But these negative feelings are actually an essential part of the process because they signal that it's time to try a new search strategy. Instead of relying on the literal associations of the left hemisphere, the brain needs to shift activity to the other side, to explore a more unexpected set of associations. It is the struggle that forces us to try something new.

What's surprising is that this mental shift often works. Because we feel frustrated, we start to look at problems from a new perspective. "You'll see people bolt up in their chair and their eyes go all wide," says Ezra Wegbreit, a graduate student in the Beeman lab who often administers the CRA test. "Sometimes, they even say 'Aha!' before they blurt out the answer." The suddenness of the insight is preceded by an equally sudden burst of brain activity. Thirty milliseconds before the answer erupts into consciousness, there's a spike of gamma-wave rhythm, which is the highest electrical frequency generated by the brain. Gamma rhythm is believed to come from the binding of neurons: cells distributed across the cortex draw themselves together into a new network that is then able to enter consciousness.

Where does this burst of gamma waves come from? To answer this question, Beeman and Kounios went back and analyzed the data from their fMRI experiment. That's when they discovered the "neural correlate of insight": the anterior superior temporal gyrus (aSTG). This small fold of tissue, located on the surface of the right hemisphere just above the ear, became unusually active in the seconds before the epiphany. (It remained silent when people solved the word puzzles by analysis.) The activation of the

cortical circuit was sudden and intense, a surge of electricity leading to a rush of blood. Although the precise function of the aSTG
remains unclear, Beeman wasn't surprised to see it involved in the
insight process. A few previous studies had linked it to aspects of
language comprehension, such as the detection of literary themes,
the interpretation of metaphors, and the comprehension of jokes.
Beeman argues that these linguistic skills share a substrate with
insight because they require the brain to make a set of distant and
original connections. Although most of us have probably never
used *age, mile,* and *sand* in a sentence before, the aSTG is able
to discover the one additional word (*stone*) that works with all of
them. And then, just when we're about to give up, the answer is
whispered into consciousness. "An insight is like finding a needle
in a haystack," Beeman says. "There are a trillion possible connections in the brain, and we have to find the exact right one. Just
think of the odds!"

Sometimes, of course, these long odds are beaten. Because
we've been stumped, we finally start searching in the correct
places, rummaging through the obscure file cabinets of the right
hemisphere. And then, if we're lucky, the search will end with a

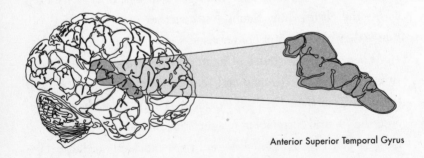

Anterior Superior Temporal Gyrus

This is the brain area that shows a spike in activity shortly
before the insight arrives.

solution, a flicker of electricity inside the head. The insight has gone incandescent.

3.

It took a few days to adjust to the quiet of Woodstock. After all, Dylan had gone straight from a frantic rock 'n' roll tour to a remote rural cabin. He was suddenly alone with nothing but an empty notebook. And there was no need to fill this notebook—Dylan had been relieved of his creative burden. For the first time in years, he didn't need to worry about his next song. Dylan told his manager that he was going to start working on a novel.

But then, just when Dylan was most determined to stop creating music, he was overcome with a strange feeling. "It's a hard thing to describe," Dylan would later remember. "It's just this sense that you got something to say." What he felt was the itch of an imminent insight, the tickle of lyrics that needed to be written down. And so Dylan did the only thing he knew how to do: he grabbed a pencil and started to scribble. Once Dylan began, his hand didn't stop moving for the next several hours. "I found myself writing this song, this story, this long piece of *vomit,* twenty pages long," Dylan said. "I'd never written anything like that before and it suddenly came to me that this is what I should do."

Vomit is the essential word here. Dylan was describing, with characteristic vividness, the uncontrollable rush of a creative insight, that flow of associations that can't be held back. "I don't know where my songs come from," Dylan said. "It's like a ghost is writing a song. It gives you the song and it goes away. You don't *know* what it means." Once the ghost arrived, all Dylan wanted to do was get out of the way.

The song that Dylan began writing in Woodstock starts like

a children's story—"Once upon a time"—but it's no fairy tale.
Dylan had no idea where this narrative was going or how it was
going to end. And so he decided to blindly follow his imagination,
as the ghost led him from one evocative image to the next:

> *Once upon a time you dressed so fine*
> *You threw the bums a dime in your prime, didn't you?*
> *People call, say, "Beware, doll, you're bound to fall."*
> *You thought they were all kiddin' you.*

What do these words mean? What is Dylan trying to tell us?
The song is an angry screed—the poetry critic Christopher Ricks
called it an "unlove song"—but who is Dylan yelling at? These
questions, of course, don't have easy answers. This was the thrilling discovery that saved Dylan's career: he could write vivid lines
filled with possibility without knowing exactly what those possibilities were. He didn't need to know. He just needed to trust the
ghost.

   This was a staggeringly strange way to create a piece of pop
music. At the time, there were two basic ways to write a song. The
first was to be like the Bob Dylan that Dylan was trying to escape:
compose serious lyrics on a serious topic. One had to sing of injustice or a broken heart, chant wordy lines over a bare-bones melody. There could be an acoustic guitar and a harmonica but not
much else.

   The second way was essentially the opposite. Instead of wal-
lowing in melancholy and complexity, one could imitate those
cynical geniuses on Tin Pan Alley and compose an irresistible jin-
gle full of major chords. Take, for instance, this number-one *Bill-
board* single from 1965, "I Can't Help Myself," made famous by
the Four Tops:

Sugar pie, honeybunch
You know that I love you

These lyrics have a deliberate clarity. As soon as the first couplet is heard, the listener knows exactly what kind of song it will be. (In this sense, it's not so different from those somber folk songs.) Such predictability is precisely what Dylan wanted to avoid; he couldn't stand the clichéd constraints of pop music. And this is why that "vomitific" writing was so important: Dylan suddenly realized that it was possible to celebrate vagueness, to write lines that didn't insist on making sense, that existed outside the categories of FM radio. He would later say that this was his first "completely free song . . . the one that opened it up for me."

But what did it open up? In retrospect, we can see that the composition—it would become the debut single on *Highway 61 Revisited*—allowed Dylan to fully express, for the first time, the diversity of his influences. Listening to these ambiguous lyrics, we can hear his mental blender at work as he effortlessly mixes together scraps of Arthur Rimbaud, Fellini, Bertolt Brecht, and Robert Johnson. There's some Delta blues and "La Bamba" but also plenty of Beat poetry, Ledbetter, and the Beatles. The song is modernist and premodern, avant-garde and country-western.

What Dylan did—and this is why he's Bob Dylan—was find the strange thread connecting those disparate voices. During those frantic first minutes of writing, his right hemisphere found a way to make something new out of this incongruous list of influences, drawing them together into a catchy song. He didn't yet know what he was doing—the ghost was still in control—but he felt the excitement of an insight, the subliminal thrill of something new. ("I don't think a song like 'Rolling Stone' could have been done any other way," Dylan insisted. "You can't sit down and write that consciously . . . What are you gonna do, chart it out?") When Dylan gets to the chorus—and he knows this is the chorus as soon as he commits it to paper—the visceral power of the song becomes obvious:

> *How does it feel*
> *To be without a home*
> *Like a complete unknown*
> *Like a rolling stone?*

The following week, on June 15, 1965, Dylan brought his sheaf of papers into the cramped space of Studio A at Columbia Records in New York City. After just four takes—the musicians were only beginning to learn their parts—"Like a Rolling Stone" was cut on acetate. Those six minutes of raw music would revolutionize rock 'n' roll. Bruce Springsteen would later describe the experience of hearing the single on AM radio as one of the most important moments of his life. Even John Lennon was in awe of the achievement.

The constant need for insights has shaped the creative process. In fact, these radical breakthroughs are so valuable that we've invented traditions and rituals that increase the probability of an epiphany, making us more likely to hear those remote associations coming from the right hemisphere. Just look at poets, who often rely on literary forms with strict requirements, such as haikus and sonnets. At first glance, this writing method makes little sense, since the creative act then becomes much more difficult. Instead of composing freely, poets frustrate themselves with structural constraints.

But that's precisely the point. Unless poets are stumped by the form, unless they are forced to look beyond the obvious associations, they'll never invent an original line. They'll be stuck with clichés and conventions, with predictable adjectives and boring verbs. And this is why poetic forms are so important. When a poet needs to find a rhyming word with exactly three syllables or an adjective that fits the iambic scheme, he ends up uncovering all sorts of unexpected connections; the difficulty of the task accelerates the insight process. Just look at Dylan's verb choice in the second stanza of "Like a Rolling Stone," which contains one of the most memorable lines in the song:

> *You've gone to the finest school all right, Miss Lonely*
> *But you know you only used to get juiced in it.*

Juiced in it? It's an incredibly effective phrase, even though the listener has no idea what it means. It's not until the next couplet that the need for *juiced* becomes clear:

> *And nobody has ever taught you how to live on the street*
> *And now you find out you're gonna have to get used to it*

Dylan uses the surprising word *juiced* because it rhymes with *used*, which is part of the snarling line that gives the stanza its literal meaning. Nevertheless, the innovative use of *juice* as a verb is one of those poetic flourishes that make "Like a Rolling Stone" so transcendent. It's a textbook example of how the imagination is unleashed by constraints. You break out of the box by stepping into shackles.

The story of "Like a Rolling Stone" is a story of creative insight. The song was invented in the moment, then hurled into the world. It took only a few seconds before a mental block became a work of art; a season of creative despair gave way to some of the most inspired music of Dylan's career. In a 1966 interview with

Playboy, Dylan assessed the impact of this sudden breakthrough on his music. "Last spring, I guess I was going to quit singing," he said. "The way things were going, it was a very draggy situation . . . But 'Like a Rolling Stone' changed it all: I didn't care anymore about writing books or poems or whatever. I mean, here was something that I myself could dig."

What Dylan dug was the strangeness of the song, the way it sounded like nothing else on the radio. In that lonely cabin, he found a way to fully express himself, to transform the fragments of art in his head into a new kind of song. He wasn't just writing a pop single — he was rewriting the possibilities of music.

2   **ALPHA WAVES (CONDITION BLUE)**

Creativity is the residue of time wasted.
—Albert Einstein

THIS IS A STORY about tape. It begins in the summer of 1925. Dick Drew was a sandpaper salesman with the Minnesota Mining and Manufacturing Company. He spent a lot of time demonstrating the effectiveness of sandpaper in auto-body shops trying to convince mechanics to buy his brand. Sometimes, after Drew made his sales pitch, he'd sit in the back of the shop and watch the men work. He soon noticed that all of the mechanics shared a common problem. It occurred when the mechanics were applying two-toned paint to a car. The workers would begin by painting everything black. Then they would protect this new coat of paint with taped-on sheets of butcher paper and carefully apply the second shade—usually a sleek line of white or red. Once the paint dried, the paper was removed. Here is where the process failed: the paper was usually attached to the metal with a strong adhesive, which meant that removing the paper and tape often peeled away the newly applied black paint. And so the frustrated workers would begin on that section again, their labor undone.

After watching this happen several times, Drew realized that the adhesive was too sticky. The workers were only hanging paper; they didn't need a strong glue. And that's when Drew had his first insight: sandpaper might help him solve the auto-body problem. Sandpaper, after all, was simply a mixture of adhesive and abrasive. (A tough paper backing coated in glue and then rolled in crushed minerals.) If you left out the abrasive, then you were left with a moderately sticky paper, which is precisely what the mechanics needed.

When Drew got back to the office after this realization, he began exploring his new idea. The first thing he discovered was that the glue used in sandpaper was also too strong—it ripped the wet paint right off. And so he began experimenting with the adhesive recipe, trying to make the rubber resin a little less sticky. This took him several months. He then had to find the right backing. Most adhesives were applied to woven fabrics, but Drew's experience as a sandpaper salesman led him to focus on a backing of paper. Unfortunately, he couldn't think of a way to store the sticky sheets; they kept sticking together, forming a crumpled stack. After two months of struggle, Drew was ordered by his boss, William McKnight, to stop working on the project. The company was in the sandpaper business; Drew should go back to selling industrial abrasive.

But Drew refused to give up. Although he was stumped, he still stayed past closing time at work, testing out different varieties of backing and recipes for glue. And then, late one night in his office, everything changed. In the time it took to have an insight—that burst of gamma waves erupting in the right hemisphere—Drew grasped the solution to his sticky problem. The idea was simple: Instead of applying the adhesive to square sheets of paper that needed to be stacked, it could be applied to a thin strip of paper that was then rolled up, like a spool of ribbon. The mechanics could unwind the necessary amount of sticky paper and attach it directly to the car, allowing them to paint without tack or glue. Drew called it masking tape.

Nobody knows where this revelation came from. Some say that Drew was inspired by the car wheels in the auto-body shop; others think he borrowed the idea from the large spools of paper that were shipped to the sandpaper factory. Drew himself had no answer. And yet, the insight happened. Drew was able to imagine a long roll of stickiness, a pressure-sensitive adhesive that could be applied to metal and then ripped off without damaging the paint.

In retrospect, the idea for a roll of tape seems incredibly obvious; it's hard to imagine a world where stickiness is limited to glue and tack and sticky sheets. Sure enough, the product was an instant hit in the marketplace, and not only among car mechanics. By 1928, Drew's company was selling more masking tape than sandpaper.

1.

The Minnesota Mining and Manufacturing Company is now called 3M. The corporate headquarters, just outside St. Paul, looks like a college campus, a sprawling five-hundred-acre landscape of lab buildings, grassy fields, and parking lots. Although the company still sells sandpaper and tape, it has since expanded into an astonishing array of product categories. (The company currently sells more than fifty-five thousand different products, giving it a nearly 1:1 product-to-employee ratio.) A random list of 3M products includes computer touch screens, kitchen sponges, water-purification filters, streetlights, stain-resistant fabrics, lithium ion batteries, home insulation, dental fillings, medical masks, and drug patches.

What do these products have in common? Nothing at all, except that they were pioneered by 3M. "We're an unusual company," says Larry Wendling, a vice president in charge of corporate research. "We have no niche or particular focus. Basically, all we do is come up with new things. It doesn't really matter what the thing is." As a result, the company spends nearly 8 percent of its gross revenue on basic research, which makes it one of the biggest spenders in the Fortune 500. While most innovative companies are celebrated for a single innovation with a short lifespan—think of Netscape, AOL, or Atari—3M has been inventing new products for more than seventy-five years. (The company was recently ranked the third most innovative company in the world, according to a survey of executives. It was beaten by Apple and Google.) Furthermore, 3M products that are less than five years old typically account for 30 percent of annual revenue, a fact that captures the constant churn of innovation at the company. "There's an astonishing diversity of research going on here," Wendling says. "I don't think there's another place that's trying to invent the next sticky tape *and* the next energy-efficient television screen *and* the next generation of vaccines. We're doing work in every scientific field."

This emphasis on innovation has been a defining feature of the company ever since Dick Drew invented masking tape. Although William McKnight, the CEO of the company at the time, initially disapproved of Drew's quixotic pursuit, he quickly saw the potential of the new product line. (Adhesives, it turned out, were much more profitable than abrasives. As one 3Mer remarked to me, "Selling tape is a great business: you make it by the mile and sell it by the inch.") And so McKnight dramatically reorganized the company, investing the tape windfall in a brand-new science lab. He hired dozens of researchers and gave them the freedom to pursue their own interests. That, after all, was the lesson of Dick Drew: even a salesman could invent an important new product.

And that's why I've come to the 3M labs in the dead of winter. I want to understand how this corporate history of innovation has informed its culture. Over the years, the company has learned a few essential tricks about creativity, and those tricks have been hard-wired into its research practices. "When you've been at this for as long as we have, you develop some important techniques for innovation," Wendling says. "We might not be the sexiest company"—he points to a wall display filled with office-supply products—"but our approach is time-tested. We know what works."

Wendling then tells me about the first essential feature of 3M innovation, which is its flexible attention policy. Instead of insisting on constant concentration—requiring every employee to focus on his or her work for eight hours a day—3M encourages people to make time for activities that at first glance might seem unproductive. Are you struggling with a difficult technical problem? Take a walk across campus. (When I visited 3M, in the late winter, the fields were full of grazing deer and employees strolling in their puffy winter parkas.) Are you stuck on a challenge that seems impossible? Lie down on a couch by a sunny window. Daydream. Play a game of pinball. While 3M demands a high level of productivity—the parking lot was full of cars at 8:00 P.M.—it also encourages employees to take regular breaks.

One important consequence of this approach was the invention of the 15 percent rule, a concept that allows every researcher to spend 15 percent of his or her workday pursuing speculative new ideas. (People at 3M refer to this time as the bootlegging hour.) The only requirement is that the researchers share their ideas with their colleagues. While bootlegging time has since been imitated at other innovative companies—Google, for instance, gives its software engineers the same freedom[1]—the concept was first implemented at 3M. "It's a little amusing that people think Google invented this idea," Wendling says. "We've been doing it here forever. At first, people thought we were crazy. They said employees need to be managed. They said the scientists would just waste their free time, that we'd be squandering all our R and D money. But here's the thing about the fifteen percent rule: it works."

The science of insight supports the 3M attention policy. Joydeep Bhattacharya, a psychologist at Goldsmiths, University of London, has used EEG to help explain why interrupting one's focus—perhaps with a walk outside or a game of Ping-Pong—can be so helpful. Interestingly, Bhattacharya has found that it's possible to predict that a person will solve an insight puzzle up to *eight seconds* before the insight actually arrives. "I never expected that we'd find such a remote precursor," he says. "It seems really strange that I can anticipate someone else's moment of insight before they are even aware of the answer. But that's what we found."

What is this predictive brain signal? The essential element is a steady rhythm of alpha waves emanating from the right hemisphere. While the precise function of alpha waves remains mys-

1. The Google program is officially known as Innovation Time Off. That program has led directly to the development of Gmail, Google's successful e-mail program, and AdSense, a nine-billion-dollar-a-year platform that allows Internet publishers to run Google ads on their sites. Marissa Mayer, Google's VP of search products and user experience, estimates that at least 50 percent of new Google products begin as Innovation Time Off speculations.

terious, they're closely associated with relaxing activities such as taking a warm shower. In fact, alpha waves are so crucial for insight that, according to Bhattacharya, subjects with insufficient alpha-wave activity are unable to utilize hints provided by the researchers. "I can give these people really obvious clues, but it still won't help," he says. "They will never get it." One of Bhattacharya's favorite insight puzzles goes like this: A man has married twenty women in a small town. All of the women are still alive and none of them are divorced. The man has broken no laws. Who is the man? Bhattacharya will let people struggle for up to three minutes before he starts giving them hints. He'll suggest possible analogies and fill his sentences with thinly veiled references to religion. However, unless the subjects are thinking in the exact right way—unless those alpha waves are visible on the EEG monitor—they will never have the insight: the man is a priest.

Why is a relaxed state of mind so important for creative insights? When our minds are at ease—when those alpha waves are rippling through the brain—we're more likely to direct the spotlight of attention *inward*, toward that stream of remote associations emanating from the right hemisphere. In contrast, when we are diligently focused, our attention tends to be directed *outward*, toward the details of the problems we're trying to solve. While this pattern of attention is necessary when solving problems analytically, it actually prevents us from detecting the connections that lead to insights. "That's why so many insights happen during warm showers," Bhattacharya says. "For many people, it's the most relaxing part of the day." It's not until we're being massaged by warm water, unable to check our e-mail, that we're finally able to hear the quiet voices in the backs of our heads telling us about the insight. The answers have been there all along—we just weren't listening.

This also helps explain the power of a positive mood. German researchers have found that when people are happy, they are much better at guessing whether or not different words share a remote associate. Even when the subjects in the German study did not find the answer—they were forced to guess after looking at words for less than two seconds—those in a positive mood were able to accurately intuit the *possibility* of an insight. In contrast, those feeling gloomy performed slightly below random chance. They had no idea which remote associates were real and which were a waste of time.

More recently, Beeman has demonstrated that people who score high on a standard measure of happiness solve about 25 percent more insight puzzles than people who are feeling angry or upset. In fact, even fleeting feelings of delight can lead to dramatic increases in creativity. After watching a short, humorous video—Beeman uses a clip of Robin Williams doing standup—subjects have significantly more epiphanies, at least when compared with those who were shown scary or boring videos. Because positive moods allow us to relax, we focus less on the troubling world and more on these remote associations. Another ideal moment for insights, according to Beeman and John Kounios, is the early morning, shortly after waking up. The drowsy brain is unwound and disorganized, open to all sorts of unconventional ideas. The right hemisphere is also unusually active. "The problem with the morning, though," Kounios says, "is that we're always so rushed. We've got to get the kids ready for school, so we leap out of bed, chug the coffee, and never give ourselves a chance to think." If you're stuck on a difficult problem, Kounios recommends setting the alarm clock a few minutes early

so that you have time to lie in bed. We do some of our best thinking when we're half asleep.[2]

One of the surprising lessons of this research is that trying to
force an insight can actually prevent the insight. While it's com-
monly assumed that the best way to solve a difficult problem is to
relentlessly focus, this clenched state of mind comes with a hidden
cost: it inhibits the sort of creative connections that lead to breakthroughs. We suppress the very type of brain activity that should
be encouraged. For instance, many stimulants taken to increase
attention, such as caffeine, Adderall, and Ritalin, seem to make
epiphanies much less likely. (According to a recent online poll
conducted by *Nature*, nearly 20 percent of scientists and researchers regularly take prescription drugs in order to improve mental
performance. The most popular reason given was "to enhance
concentration.") Because these stimulants shift attention away
from the networks of the right hemisphere, they cause people to
ignore those neurons that might provide the solution. "People assume that increased focus is always better," says Martha Farah, a
neuroscientist at the University of Pennsylvania. "But what they
don't realize is that intense focus comes with real tradeoffs. You
might be able to work for eight hours straight [on these drugs], but
you're probably not going to have many big insights."[3]

....

2. *There's one additional cortical signal that predicts epiphanies. Looking at the data,
Beeman and Kounios saw a sharp drop in activity in the visual cortex just before the
insight appeared, as if the sensory area were turning itself off. At first, the scientists
couldn't figure out what was going on. But as they were struggling to decipher the data,
Beeman watched Kounios cover his eyes with his hand. That's when it occurred to him:
the visual cortex was going quiet so that the brain could better focus on its own obscure
associations. "The cortex does this for the same reason we close or cover our eyes when
we're trying to think," Beeman says. When the outside world becomes distracting, the
brain automatically blocks it out.*
3. *Marijuana, by contrast, seems to make insights more likely. It not only leads to states
of relaxation but also increases brain activity in the right hemisphere. A recent paper
by scientists at University College, London, looked at a phenomenon called semantic*

Consider an experiment that investigated the problem-solving abilities of neurological patients with severe attention problems. (Most of these patients had suffered damage to the prefrontal cortex, a part of the brain just behind the forehead.) Because of their injuries, these poor people lived in a world of endless distractions; their focus was always fleeting. Here's a sample problem given to the brain-damaged patients:

IV = III + III

The task is to move a single line so that the false arithmetic statement becomes true. (In this example, you would move the first *I* to the right side of the *V* so that it reads VI = III + III.) Nearly 90 percent of the brain-damaged patients were able to correctly solve the puzzle, since it required a fairly obvious problem-solving approach: the only thing you have to do is change the answer. (A group of subjects without any attention deficits found the answer 92 percent of the time.) But here's a much more challenging equation to fix:

III = III + III

In this case, only 43 percent of normal subjects were able to solve the problem. Most stared at the Roman numerals for a few

....

priming. *This occurs when the activation of one word allows an individual to react more quickly to related words. For instance, the word* dog *might lead to faster reaction times for* wolf, pet, *and* Lassie, *but it won't alter how quickly a person reacts to* chair.

Interestingly, the scientists found that marijuana seems to induce a state of hyperpriming, meaning that it extends the reach of semantic priming to distantly related concepts. As a result, one hears dog *and thinks of nouns that in more sober circumstances would seem completely disconnected. This state of hyperpriming helps explain why cannabis has so often been used as a creative fuel: it seems to make the brain better at detecting the remote associations that define the insight process.*

minutes and then surrendered. The patients who couldn't pay attention, however, had an 82 percent success rate. This bizarre result—brain damage leads to dramatically *improved* performance—has to do with the unexpected nature of the solution: rotate the vertical line in the plus sign by ninety degrees, transforming it into an equal sign. (The equation is now a simple tautology: III = III = III.) The reason this puzzle is so difficult, at least for people without brain damage, has to do with the standard constraints of math problems. People are not used to thinking about the operator in an equation, so most of them quickly fix their attention on the Roman numerals. But that's a dead end. The patients with severe cognitive deficits, by contrast, can't restrict their search. They are forced by the brain injury to consider a much wider range of possible answers. And this is why they're nearly twice as likely to have an insight.

Or look at a recent study led by Holly White, a psychologist at the University of Memphis. White began by giving a large sample of undergraduates a variety of difficult creative tests. Surprisingly, those students diagnosed with attention deficit hyperactivity disorder (ADHD) got significantly higher scores. White then measured levels of creative achievement in the real world, asking the students if they'd ever won prizes at juried art shows or been honored at science fairs. In every single domain, from drama to engineering, the students with ADHD had achieved more. Their attention deficit turned out to be a creative blessing.

The unexpected benefits of *not* being able to focus reveal something important about creativity. Although we live in an age that worships attention—when we need to work, we force ourselves to concentrate—this approach can inhibit the imagination.

Sometimes it helps to consider irrelevant information, to eavesdrop on all the stray associations unfolding in the far reaches of the brain. Occasionally, focus can backfire and make us fixated on the wrong answers. It's not until you let yourself relax and indulge in distractions that you discover the answer; the insight arrives only after you stop looking for it.

Kounios tells a story about a Zen Buddhist meditator that illustrates the importance of these alpha waves. At first, this man couldn't solve any of the CRA problems given to him by the scientists. "This guy went through thirty or so of the verbal puzzles and just drew a blank," Kounios says. "He assumed the way to solve the problems was to think really hard about the words on the page, to really concentrate." But then, just as the meditator was about to give up, he started solving one puzzle after another; by the end of the experiment, he was getting them all right. It was an unprecedented streak. According to Kounios, this dramatic improvement depended on the ability of the meditator to focus on not being focused so that he could finally pay attention to all those fleeting connections in the right hemisphere. "Because he meditated ten hours a day, he had the cognitive control to instantly relax," Kounios says. "He could ramp up those alpha waves at will, so that all of a sudden he wasn't paying such close attention to the words on the page. And that's when he became an insight machine."

2.

While 3M's flexible attention policy is a pillar of its innovation culture, the company doesn't rely on relaxation and distraction alone to generate new insights. As Wendling notes, "Sometimes, you've got to take a more active role . . . We want to give our researchers freedom, but we also want to make sure the ideas they're pursuing are really new and worthwhile." This is where horizontal sharing, the second essential feature of the 3M workplace, comes in. The idea is rooted in the company's tradition of inventing new products by transplanting the same concept into different domains. Just consider the invention of masking tape. Drew's fundamental insight was that even a simple product like sandpaper — nothing but sturdy paper coated with a sticky glue — could have multiple uses; Drew realized that those ingredients could also be turned into a roll of adhesive. This led William McKnight, the executive who turned 3M into an industrial powerhouse, to insist on sharing among scientists as a core tenet of 3M culture. Before long, the Tech Forum was established, an annual event at which every researcher on staff presents his or her latest research. (This practice has also been widely imitated. Google, for instance, hosts a conference called CSI, or Crazy Search Ideas.) "It's like a huge middle-school science fair," Wendling says. "You see hundreds of posters from every conceivable field. The guys doing nanotechnology are talking to the guys making glue. I can only imagine what they find to talk about."

The benefit of such horizontal interactions — people sharing knowledge *across* fields — is that it encourages conceptual blending, which is an extremely important part of the insight process. Normally, the brain files away ideas in categories based on how these ideas can be used. If you're working for a sandpaper company, for instance, then you probably spend most of the day thinking about sandpaper as an abrasive. That, after all, is the purpose of the product. The assumption is that the vast store of mental concepts work only in particular situations and that it's a waste of time to apply them elsewhere. There's no point in thinking about sandpaper if you don't need to sand something down.

Most of the time, this assumption holds true. However, the

same tendency that keeps us from contemplating irrelevant concepts also keeps us from coming up with insights. The reason is that our breakthroughs often arrive when we apply old solutions to new situations; for instance, a person thinking about sandpaper when he needs something sticky. Instead of keeping concepts separate, we start blending them together, trespassing on the standard boundaries of thought.

The best way to understand conceptual blending is to look at the classic children's book *Harold and the Purple Crayon*. The premise of the book is simple: Harold has a magic crayon. When he draws with this purple crayon, the drawing becomes real, although it's still identifiable as a childish sketch. If Harold wants to go for a walk, he simply draws a path with his crayon. This fictive sketch then transforms into a real walkway, which Harold can stroll along. This magic crayon is seemingly the solution to every problem.

But here's the twist that makes *Harold and the Purple Crayon* such an engaging book: it blends together two distinct concepts of the world. Although the magic crayon is clearly a fantastical invention—a conceit that could never exist—Harold still has to obey the rules of reality. So when Harold draws a mountain and then climbs it, he must try not to slip and fall down. When he does slip—gravity exists even in this crayon universe—Harold has to draw a balloon to save himself. In other words, the book is delicate blend of the familiar and the fictional; Harold has a surreal tool, but it operates amid the usual constraints. Mark Turner, a cognitive psychologist at Case Western Reserve University, has used this children's book to demonstrate that even little kids can easily combine two completely distinct concepts into a single idea. If they couldn't, then the travails of Harold would make no sense.

What does conceptual blending have to do with creativity?

Although people take this mental skill for granted, the ability to make separate ideas coexist in the mind is a crucial creative tool. Insights, after all, come from the overlap between seemingly unrelated thoughts. They emerge when concepts are transposed, when the rules of one place are shifted to a new domain. The eighteenth-century philosopher David Hume, in *An Enquiry Concerning Human Understanding*, described this talent as the essence of the imagination:

> All this creative power of the mind amounts to no more than the faculty of compounding, transposing, augmenting, or diminishing the materials afforded us by the senses and experience. When we think of a golden mountain, we only join two consistent ideas, *gold,* and *mountain,* with which we were formerly acquainted.

Hume was pointing out that the act of invention was really an act of recombination. The history of innovation is full of inventors engaged in "compounding" and "transposing." Johannes Gutenberg transformed his knowledge of winepresses into an idea for a printing machine capable of mass-producing words. The Wright brothers used their knowledge of bicycle manufacturing to invent the airplane. (Their first flying craft was, in many respects, just a bicycle with wings.) George de Mestral came up with Velcro after noticing burrs clinging to the fur of his dog. And Larry Page and Sergey Brin developed the search algorithm behind Google by applying the ranking method used for academic articles to the sprawl of the World Wide Web; a hyperlink was like a citation. In each case, the radical concept was merely a new mixture of old ideas.

Dick Drew was a master at conceptual blending. After he invented masking tape, a colleague told him about a strange new material called cellophane. (By this time, Drew had become a full-time researcher.) The material was translucent and shiny but also strikingly impermeable to water and grease; it was being sold by DuPont as a packaging solution, a cheap way of wrapping products for shipping. Drew took one look at the material and had another idea, which he would later describe as the insight of his life: cellophane would make a perfect adhesive. He ordered a hundred yards of cellophane and began coating the material with glue. Drew called it Scotch tape. By 1933, less than two years after the see-through adhesive hit the market, the product had become the most popular consumer tape in the world. Although masking tape and cellophane were completely unrelated—it had never occurred to DuPont researchers to make their wrapping material sticky—Drew saw their possible point of intersection.

This process has been repeated again and again at 3M. For instance, the adhesive used in industrial-strength masking tape gave rise to the sound-dampening panels used in Boeing aircraft. (The material is so sticky that it even binds sound waves.) Those panels in turn gave rise to the extremely strong adhesive foam used in golf clubs, which can hold together carbon fiber and titanium during high impact. And the concept of Scotch tape eventually inspired another 3M engineer to invent the touch-screen technology used in smartphones. (Instead of coating cellophane, the clear glue is used to coat an electrically charged glass surface, which is then attached to a display.) After a 3M engineer noticed that Scotch tape could act like a prism, a team of scientists used their tape expertise to develop transparent films that refract light. Such films are now being widely used in laptops and LCD televisions; because they direct the brightness of each bulb outward, fewer bulbs are required on the inside, thus reducing the energy consumption of the devices by as much as 40 percent. "The lesson is that the tape business isn't just about tape," Wendling says. "You might think an idea is finished, that there's nothing else to do with it, but then you talk to somebody else in some other field. And your little idea inspires them, so they come up with a brand-new invention that inspires someone else. That, in a nutshell, is our model."

In fact, 3M takes conceptual blending so seriously that it regularly rotates its engineers, moving them from division to division. A scientist studying adhesives might be transferred to the optical-films department; a researcher working on asthma inhalers might end up tinkering with air conditioners. Sometimes, these rotations are used as a sudden spur for innovation. If a product line is suffering from a shortage of new ideas, 3M will often bring in an entirely new team of engineers, sourced from all over the company. "Our goal is to have people switch problems every four to six years," Wendling says. "We want to ensure that our good ideas are always circulating."

The benefit of such circulation is that it increases conceptual blending, allowing people to look at their most frustrating problems from a fresh perspective. Instead of trying to invent a new tack, imagine a roll of sticky paper; instead of trying to improve the battery performance of a laptop, think about the refractory properties of its light bulbs. To get a better sense of how this mental process unfolds, consider this insight puzzle, which is notoriously difficult:

> You are a doctor faced with a patient who has a malignant tumor in his stomach. It is impossible to operate on the patient, but unless the tumor is destroyed, the patient will die. There is a kind of ray machine that can be used to shoot at and destroy the tumor. If the rays reach the tumor all at once at a sufficiently high intensity, the tumor will be destroyed. Unfortunately, at this in-

tensity, the healthy tissue that the rays pass through on the way to the tumor will also be destroyed. At lower intensities the rays are harmless to healthy tissue, but they will not affect the tumor either. What type of procedure might be used to destroy the tumor with the rays, and at the same time avoid destroying the healthy tissue?

If you can't figure out the answer, don't worry; more than 97 percent of people conclude that the problem is impossible — the patient is doomed. However, there's a very simple way to dramatically boost the success rate of solving this insight puzzle. It involves telling the subjects a story that seems entirely unrelated:

> A fortress was located in the center of the country. Many roads radiated out from the fortress. A general wanted to capture the fortress with his army. But he also wanted to prevent mines on the roads from destroying his army and neighboring villages. As a result, the entire army could not all go down one road to attack the fortress. However, the entire army was needed to capture the fortress; an attack by one small group could not succeed. The general therefore divided his army into several small groups. He positioned the small groups at equal distances from the fortress along different roads. The small groups simultaneously converged on the fortress. In this way the army captured the fortress.

When the tumor puzzle was preceded by this military tale, nearly 70 percent of subjects came up with the solution. Because the subjects were able to see what the different stories had in common, they generated a moment of insight; the answer emerged from the analogy. (If you are still wondering, the solution to the doctor's problem is to mount ten separate ray guns around the patient and set each of them to deliver 10 percent of the necessary radiation. When the ray machines are all focused on the stomach, there is enough radiation to destroy the tumor while preserving the surrounding tissue.)

How can we get better at conceptual blending? According to Mary Gick and Keith Holyoak, the psychologists behind the tumor puzzle, the key element is a willingness to consider information and ideas that don't seem worth considering. Instead of concentrating on the details of the problem—most people quickly fixate on tumors and rays—we should free our minds to search for distantly related analogies that can then be mapped onto the puzzles we're trying to solve. Sometimes, the best way to decipher a medical mystery is to think about military history.

The importance of considering the irrelevant helps explain a recent study led by neuroscientists at Harvard and the University of Toronto. The researchers began by giving a sensory test to eighty-six Harvard undergraduates. The test was designed to measure their ability to ignore outside stimuli, such as the air conditioner humming in the background or the conversation taking place in a nearby cubicle. This skill is typically seen as an essential component of productivity, since it keeps people from getting distracted by extraneous information. Their attention is less likely to break down.

Here's where the data get interesting: those undergrads who had a tougher time ignoring unrelated stuff were also *seven* times more likely to be rated as "eminent creative achievers" based on their previous accomplishments. (The association was particularly strong among distractible students with high IQs.) According to the scientists, the inability to focus helps ensure a richer mixture of thoughts in consciousness. Because these people had difficulty filtering out the world, they ended up letting more in. Instead of approaching the problem from a predictable perspective, they considered all sorts of far-fetched analogies, some of which proved useful.[4] "Creative individuals seem to remain in contact with the extra information constantly streaming in from the environment," says Jordan Peterson, a neuroscientist at the University of Toronto and lead author on the paper. "The normal person classifies an object, and then forgets about it. The creative person, by contrast, is always open to new possibilities."

3.

Marcus Raichle, a neurologist and radiologist at Washington University, got interested in daydreaming by accident. It was the early 1990s, and Raichle was studying the rudiments of visual perception. His experiments were straightforward: A subject performed a particular task, such as counting a collection of dots, in a brain scanner. Then he or she did nothing for thirty seconds. ("It was pretty boring for the subjects," Raichle admits. "You always had to make sure people weren't dozing off.") Although the scanner was still collecting data in between the actual experiments, Raichle assumed that this information was worthless noise. "We told the subjects to not think about anything," he says. "We wanted them to have a blank mind. I assumed that this would lead to a real drop in brain activity. But I was wrong."

One day, Raichle decided to analyze the fMRI data collected when the subjects were just lying in the scanner waiting for the

---

4. *Another useful trick for inciting insights involves a quirk of language. According to an experiment led by Catherine Clement at Eastern Kentucky University, one way to consistently increase problem-solving ability is to change the verbs used to describe the problem. When the verbs are extremely specific, creativity is constrained, and people struggle to find useful comparisons. However, when the same problem is recast with more generic verbs, people are suddenly more likely to uncover unexpected parallels. In some instances, Clement found, the simple act of rewriting the problem led to stunning improvements in the performance of her subjects. Insight puzzles that had seemed impossible—not a single person was able to solve them—were now solved more than 60 percent of the time.*

next task. (He needed a baseline of activity.) To his surprise, Raichle discovered that the brains of subjects were not quiet or subdued. Instead, they were overflowing with thoughts, their cortices lit up like skyscrapers at night. "When you don't use a muscle, that muscle isn't doing much," Raichle says. "But when your brain is supposedly doing nothing, it's really doing a tremendous amount."

Raichle was fascinated by the surge in brain activity between tasks. At first, he couldn't figure out what was happening. But while sitting in his lab one afternoon, he came up with the answer: The subjects were daydreaming! ("I was probably daydreaming when the idea came to me," Raichle says.) Because they were bored silly in the claustrophobic scanner, they were forced to entertain themselves. This insight immediately led Raichle to ask the next obvious question: Why did daydreaming consume so much energy? "The brain is a very efficient machine," he says. "I knew that there must be a good reason for all this neural activity. I just didn't know what the reason was."

After several years of patient empiricism, Raichle began outlining a mental system that he called the default network, since it appears to be the default mode of thought. (We're an absentminded species, constantly disappearing down mental rabbit holes.) This network is most engaged when a person is performing a task that requires little conscious attention, such as routine driving on the highway or reading a tedious book. People had previously assumed that daydreaming was a lazy mental process, but Raichle's fMRI studies demonstrated that the brain is extremely busy during the default state. There seems to be a particularly elaborate electrical conversation between the front and back parts of the brain, with the prefrontal folds (located just behind the eyes) firing in sync with the posterior cingulate, medial temporal lobe, and precuneus. These cortical areas don't normally interact directly; they have different functions and are part of distinct neural pathways. It's not until we start to daydream that they begin to work closely together.

All this mental activity comes with a very particular purpose. Instead of responding to the outside world, the brain starts to explore its inner database, searching for relationships in a more relaxed fashion. (This mental process often runs parallel with increased activity in the right hemisphere.) Virginia Woolf, in her novel *To the Lighthouse,* eloquently describes this form of thinking as it unfolds inside the mind of a character named Lily:

> Certainly she was losing consciousness of the outer things. And as she lost consciousness of outer things . . . her mind kept throwing things up from its depths, scenes, and names, and sayings, and memories and ideas, like a fountain spurting . . .

A daydream is that "fountain spurting" as the brain blends together concepts that are normally filed away in different areas. The result is an ability to notice new connections, to see the overlaps that we normally overlook. Take, for instance, the story of Arthur Fry, an engineer at 3M in the paper-products division. It begins on a frigid Sunday morning in 1974 in the front pews of a Presbyterian church in north St. Paul, Minnesota. A few weeks earlier, Fry had attended a Tech Forum presentation by Spencer Silver, an engineer working on—you guessed it—adhesives. Silver had developed an extremely weak glue, a paste so feeble it could barely hold two pieces of paper together. Like everyone else in the room, Fry had patiently listened to the presentation and then failed to come up with any practical applications for the compound. "It seemed like a dead-end idea," Fry says. "I quickly put it out of my thoughts." What good, after all, is a glue that doesn't stick?

That Sunday, however, the paste reentered Fry's thoughts, albeit in a rather unlikely context. "I sang in the church choir," Fry remembers, "and I would often put little pieces of paper into the music on Wednesday night to mark where we were singing. Sometimes, before Sunday morning, those little papers would fall out." This annoyed Fry, because it meant that he would often spend the service frantically thumbing through his hymnal, looking for the right page. But then, during a particularly boring sermon, Fry engaged in a little daydreaming. He began thinking about bookmarks, and how what he needed was a bookmark that would stick to the paper but wouldn't tear it when it was removed. And that's when Fry remembered Spencer Silver and his ineffective glue. He immediately realized that Silver's patented formula—this barely sticky adhesive—could help create the perfect bookmark.

So Fry started working, in his bootlegging time, on this new product for his hymnal. After several months of chemical tinkering—the first bookmarks destroyed his books, leaving behind a gluey residue—Fry developed a working prototype, which became the basis for a small test run. "I gave some of them to my cohorts in the lab, to secretaries, to the librarians," he says. "Basically anybody who would take them." Although people found the product useful—it was better than folding down page corners—nobody wanted a refill. Instead of disposing of the bookmarks, Fry's coworkers just transferred them from book to book.

Fry was ready to give up. But then, a few weeks later, Fry had a second epiphany. He was reading a report and had a question about a specific paragraph. However, instead of writing a note directly on the paper, Fry cut out a square of the bookmark material, stuck it onto the page, and wrote his question there. He sent the report to his supervisor; Fry's supervisor jotted down his response on a different sticky square, applied it to another document, and sent that back. The men immediately realized they'd discovered a new way to communicate. Instead of writing separate memos full of page references and excerpted quotes, they could stick questions and comments directly onto the text. And these sticky little papers weren't useful just for documents; every surface in the office was now a potential bulletin board. This time when Fry gave out the products to colleagues, he suggested that they write on them. Within weeks, the 3M offices were plastered with canary-yellow squares. The Post-it note was born.

It's not an accident that Arthur Fry was daydreaming when he came up with the idea for a sticky bookmark. A more disciplined thought process wouldn't have found the connection between Spencer Silver's weak adhesive and the annoying tendency of those pieces of paper to fall out of the choral book. The errant daydream is what made Post-it notes possible. The boring sermon didn't hurt either.

Jonathan Schooler, the psychologist who helped pioneer the study of insight, has recently begun studying the benefits of daydreams. His lab has demonstrated that people who consistently engage in more daydreaming score significantly higher on measures of creativity. (To evaluate daydreaming, he gave subjects a slow section of *War and Peace* and then timed how long it took them to start thinking about something else.) "What these tests measure is someone's ability to find hidden relationships that can help them solve a problem," Schooler says. "That kind of thinking is the essence of creativity. And it turns out that people who daydream a lot are much better at it."

However, not all daydreams are equally effective at inspiring useful new ideas. In his experiments, Schooler distinguishes between two types of daydreaming. The first type occurs when people notice they are daydreaming only when prodded by the researcher. Although they've been told to press a button as soon as they realize their minds have started to wander, these people fail to press the button. The second type of daydreaming occurs when people catch themselves during the experiment—they notice they're daydreaming on their own. According to Schooler's data, individuals who are unaware that their minds have started wandering don't exhibit increased creativity. "The point is that it's not enough to just daydream," Schooler says. "Letting your mind drift off is the easy part. The hard part is maintaining enough awareness so that even when you start to daydream you can interrupt yourself and notice a creative thought." In other words, the reason Fry is such a good inventor—he has more than twenty patents to his name, in addition to Post-it notes—isn't simply that he's a prolific mind-wanderer. It's that he's able to pay attention to his daydreams and to detect those moments when his daydreams generate insights.

This helps explain another interesting experiment from Schooler's lab, described in a 2009 paper wittily entitled "Lost in the Sauce." In this study, Schooler once again had undergraduates read a boring passage from *War and Peace*. However, he first gave some of the students a generous serving of vodka and cranberry juice. Not surprisingly, the drunk readers had a tougher time paying attention to the text than their sober classmates did and were much more likely to engage in idle daydreams. More important, though, the students given alcohol almost always failed to notice they had stopped paying attention to Tolstoy until probed. Schooler suggests that's because alcohol induces a particularly intense state of mind-wandering, which he refers to as zoning out. "This is why it's nice to have a beer or two after work," Schooler says. "We become a little less aware of what we're thinking. But that awareness is also the key to a productive session of mind-wandering. You might solve a problem while drunk, but you probably won't notice the answer."

The lesson is that productive daydreaming requires a delicate mental balancing act. On the one hand, translating boredom into a relaxed form of thinking leads to a thought process characterized by unexpected connections; a moment of monotony can become a rich source of insights. On the other hand, letting the mind wander so far away that it gets lost isn't useful; even in the midst of an entertaining daydream, you need to maintain a foothold in the real world.

Schooler has begun applying this research to his own life: he now takes a dedicated daydreaming walk every day. He shows me his favorite route, a hiking trail on the bluffs above a scenic Santa Barbara beach. The landscape is chaparral and oak trees; the only sound is the rhythm of the waves below. "This is where I come to relax," Schooler says. "But just because I'm relaxed doesn't mean I'm not working. What I realized is that the kind of thinking I do here [on the hike] is so useful that I needed to build it into my work routine. It wasn't enough to just daydream in my spare moments, while sitting in traffic or waiting in line. I needed to be more disciplined about my mind-wandering." And so, every afternoon, Schooler parks his car on the Pacific Coast Highway, leaves his iPhone behind, and walks along these seaside cliffs. "I never have a plan or a list of things I need to think about," he says. "Instead, I just let my mind go wherever it wants. And you know what? This is where I have all my best ideas."

The advantage of knowing where insights come from is that it can make it easier to generate insights in the first place. When we're struggling with seemingly impossible problems, it's important to find time to unwind, to eavesdrop on all those remote associations coming from the right hemisphere. Instead of drinking another cup of coffee, indulge in a little daydreaming. Rather than relentlessly focusing, take a warm shower, or play some Ping-Pong, or walk on the beach.

Look at this recent experiment, published in *Science*. These psychologists, at the University of British Columbia, were interested in looking at how various colors influence the imagination. They recruited six hundred subjects, most of them undergraduates, and had them perform a variety of basic cognitive tests displayed against red, blue, or neutral backgrounds.

The differences were striking. When people took tests in the red condition, they were much better at skills that required accuracy and attention to detail, such as catching spelling mistakes or keeping random numbers in short-term memory. According to the scientists, this is because people automatically associate red with danger, which makes them more alert and aware.

The color blue, however, carried a completely different set of psychological benefits. While people in the blue group performed worse on short-term memory tasks, they did far better on those requiring some imagination, such as coming up with creative uses for a brick or designing a children's toy out of simple geometric shapes. In fact, subjects in the blue condition generated *twice* as many creative outputs as did subjects in the red condition.[5]

We can now begin to understand why being surrounded by blue walls makes us more creative. According to the scientists,

. . . .

5. *Interestingly, exposing subjects to an incandescent light bulb can also increase performance on a variety of insight puzzles. Because the lit bulb is a cliché of insight, the cheap cultural artifact makes people more sensitive to those quiet insights coming from the right hemisphere.*

the color automatically triggers associations with the sky and ocean. We think about expansive horizons and diffuse light, sandy beaches and lazy summer days; alpha waves instantly increase.[6] This sort of mental relaxation makes it easier to daydream and pay attention to insights; we're less focused on what's right in front of us and more aware of the possibilities simmering in our imaginations.

There's something deeply surprising about these data. We tend to assume that some people are simply more creative than others, that originality is a predetermined personality trait: if a person isn't born with the correct kind of brain, he'll never be able to compose an original song or come up with an idea as innovative as Post-it notes. But creativity isn't a fixed feature of the mind—that's why merely exposing people to the color blue can double their creative output. The imagination is vaster than we can imagine. We just need to learn how to listen.

---

6. *And it's not just blueness; the scientists speculate that any open, sunny space can lead to increased creativity. Architecture has real cognitive consequences.*

3 THE UNCONCEALING

It was a flash of inspiration. Kind of a thirty-year flash.
— Charles Eames

THE POET W. H. AUDEN was a drug addict. He began every day the same way: He sat at his cluttered desk and gulped a cup of strong black coffee. Then he smoked a cigarette. Auden needed the caffeine and nicotine—"I need them quite desperately," he admitted—but they were never enough. And so, before he began to write, Auden took a little white pill of Benzedrine, an amphetamine that accelerated his brain. After just a single dose, he was able to think with astonishing speed, the intricate poetry pouring onto the sheets of blank paper. "The drug is a labor-saving device," Auden said. "It turns me into a working machine."

Auden was introduced to Benzedrine in 1938 by an American editor while he was on a short vacation in New York City. The drug wasn't illegal—it had been marketed since 1927 as a treatment for asthma—and was regularly prescribed by doctors for a wide variety of other ailments as well, including obesity, impotence, and migraines. Although Auden began taking the amphetamine to stay awake—he wanted to explore Manhattan night-clubs—he quickly realized that the drug was a powerful writing tool that helped him concentrate for hours at a time on the details of his poetry. Writing verse was an exhausting process for him, but the pills allowed him to persist, to play with his words until they were perfect. (When he was done working, Auden would wash away the Benzedrine with a martini and barbiturates.) He liked to quote Valéry on the subject: "A person is a poet if his imagination is *stimulated* by the difficulties inherent in his art and not if his imagination is dulled by them." Auden was stimulated by those difficulties, but it still helped to take some stimulants.

Auden wasn't the only literary speed addict; writers and amphetamines have a rich and tangled history. Robert Louis Stevenson wrote *The Strange Case of Dr. Jekyll and Mr. Hyde* during a six-day cocaine binge. Graham Greene, James Agee, and Philip K. Dick were all Benzedrine addicts; they treated the drug like a multivitamin for the mind. Dick, for instance, once remarked that his work could be classified into two categories: "The writing that was done under the influence of drugs and the writing I've done when I'm not under the influence of drugs. But when I'm not under the influence of drugs what I do is I write about drugs." Like Auden, Dick started taking Benzedrine to fulfill a practical need: he needed to stay awake to write. "I had to write so much in order to make a living because our pay rates were so low," he said. "The amphetamines gave me the energy." While addicted to Benzedrine, Dick completed sixteen novels in five years.

Or look at Jack Kerouac: he steeped his brain in Benzedrine so that he could write *On the Road* in an epic three-week "kick-writing" session, working eighteen-hour days hunched over his tiny typewriter, a scroll of telegraph paper fed into the machine. The result was a first draft of the Beat novel. While Kerouac would later edit the prose—the 120-foot-long scroll is crammed with marks and revisions—the sense of breathless speed remained. You can hear the amphetamine in the run-on sentences.

And it wasn't just writers who abused Benzedrine. Paul Erdos, one of the most productive mathematicians of the twentieth century, was a notorious amphetamine addict. (He coauthored 1,475 peer-reviewed-journal articles over the course of his career, or nearly a paper every two weeks.) Erdos, a skinny man with oversize glasses, lived out of a single suitcase for most of his life. He seemed to subsist on little more than cookies and caffeine tablets and is said to have remarked that "a mathematician is a machine for turning coffee into theorems."[1] Erdos supplemented those uppers with amphetamines, and when he was absorbed in a particularly difficult problem he was prone to bouts of weight loss. Ron Graham, a friend and fellow mathematician, once bet Erdos five hundred dollars that he couldn't abstain from amphetamines for thirty days. Erdos won the wager but complained that the progress of mathematics had been set back by a month: "Before, when I looked at a piece of blank paper my mind was filled with ideas," he complained. "Now all I see is a blank piece of paper."

At first glance, this kind of focused thought process does not seem very imaginative. Creativity is typically associated with epiphanies, not struggling to stay focused. We like to talk about our revelations, not the painstaking work that surrounds them. Although people value grit and persistence—or at least we say we do—we don't *admire* these traits, at least not in the same way

1. *Coffee is a stimulant, like a very mild version of Benzedrine. Bennett Alan Weinberg and Bonnie K. Bealer, in* The World of Caffeine, *argue that the spread of caffeinated beverages in the seventeenth century made the Industrial Revolution possible, since it allowed "large numbers of people to coordinate their work schedules by giving them the energy to start work at a given time and continue it as long as necessary." Before coffee and tea were widely available, the European breakfast drink of choice had been beer, so the psychological contrast was especially dramatic.*

that we admire acts of radical genius. In fact, most of us see perseverance as a distinctly uncreative approach, the sort of strategy that people with mediocre ideas are forced to rely on. And this is why it's easy to disregard Auden as just a doped-up copyeditor or dismiss Jack Kerouac as a second-rate writer. (After reading *On the Road,* Truman Capote remarked, "That's not writing—that's *typing.*")

But Capote was wrong. The reality of the creative process is that it often requires persistence, the ability to stare at a problem until it makes sense. It's forcing oneself to pay attention, to write all night and then fix those words in the morning. It's sticking with a poem until it's perfect; refusing to quit on a math question; working until the cut of a dress is just right. The answer won't arrive suddenly, in a flash of insight. Instead, it will be revealed slowly, gradually emerging after great effort.

The larger point is that creativity isn't just about relaxing showers and remote associations. That's how Dylan wrote "Like a Rolling Stone," but that's not the only way to make something new. The imagination, it turns out, is multifaceted. And so, when the right hemisphere has nothing to say, when distractions are just distractions, we need to rely on a very different circuit of cells. We can't always wait for the insights to find us; sometimes, we have to search for them.

Furthermore, even if a person is lucky enough to experience a useful epiphany, that new idea is rarely the end of the creative process. The sobering reality is that the grandest revelations often still need work. The new idea—that thirty-millisecond burst of gamma waves—has to be refined, the rough drafts of the right hemisphere transformed into a finished piece of work. Such labor is rarely fun, but it's essential. A good poem is never easy. It must be pulled out of us, like a splinter.

What does this have to do with Benzedrine? The answer returns us to the brain and to the specific ways in which amphetamines alter the activity of neurons. After Auden popped his morning pill, the drug quickly diffused into his bloodstream. Most chemicals can't pass through the blood-brain barrier—a mostly impermeable wall that protects the mind from mind-altering substances—but the small molecules of speed slip right through it. As a result, Auden began feeling the effects of Benzedrine within five minutes of swallowing the white tablet.

While the drug might have turned the poet into a poetry machine, it came with many dangerous side effects. For one thing, Benzedrine is extremely addictive. It's also been known to cause insomnia, psychotic episodes, tremors, constipation, and cardiac arrest. Furthermore, such stimulants can block moments of insight. Because the drugs sharpen the spotlight of attention, they make it much harder for anyone to hear those remote associations emanating from the right hemisphere. The distracting murmurs of the mind are silenced; the alpha waves disappear.

However, even though these stimulants inhibit our epiphanies (and sicken us with addiction), they seem to dramatically increase *other* kinds of creativity. Just look at Auden: it's astonishing how many of his most celebrated poems, from "Musée des Beaux Arts" to "In Memory of W. B. Yeats," were composed in those first insomniac months after he started experimenting with Benzedrine. In fact, Auden's poetic production during this time is widely considered to be one of the great outpourings of literature in the twentieth century. It's as if the pills awakened his latent talent, transforming a gifted lyricist into the premier poet of his generation.

But the amphetamines didn't just make Auden insanely productive—they also transformed his writing style. His flood of

Benzedrine poetry is defined by its clarity, the lines stripped down to their rhetorical essence. There are no wild rhymes here—every word is deliberate. Take this stanza from "September 1, 1939," written in the dark days right before World War II:

All I have is a voice
To undo the folded lie
The romantic lie in the brain
Of the sensual man-in-the-street
And the lie of Authority
Whose buildings grope the sky:
There is no such thing as the State
And no one exists alone;
Hunger allows no choice
To the citizen or the police;
We must love one another or die.

The writing feels candid, as if it were composed on the back of a cocktail napkin. (The poem opens, after all, with a bar scene: "I sit in one of the dives / On Fifty-Second Street.") But that ease is an illusion: Auden lavished months of attention on these lyrics, patiently fixing the flaws and cutting the excess. For the most part, his drafts document a move toward simplicity, so that the more opaque lines—"Each pert philosopher's / Concupiscence or, worse / Practical wisdom, all"—were removed. The power of the finished poem is inseparable from this amphetamine-fueled editing process: for the first time, Auden was fully able to focus on the writing until it was lean and spare and ready for publication. He no longer got bored with bad metaphors or stopped working on substandard stanzas. Instead, the drugs allowed him to relentlessly refine his words. He thought about the lines, and then he thought some more.

But the question remains: How does a little white pill make this kind of focused creativity easier? Why does Benzedrine make someone more likely to persevere? At first glance, the effects of the drug on the brain seem relatively minor. Amphetamines act primarily on a network of neurons that use dopamine, a neurotransmitter, to communicate with one another. (Our cells speak in squirts of chemical.) Within minutes, the drug dramatically increases the amount of dopamine in the synapses, which are the spaces between cells.[2] This excess of neurotransmitter means that neurons are stuck in the active state, like a light that can't be turned off.

While dopamine neurons are relatively rare, they are clustered in specific areas in the center of the brain, such as the nucleus ac-

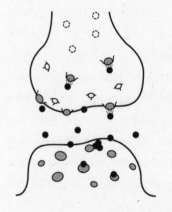 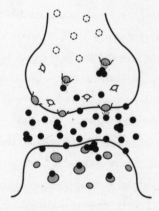

Normal Release of Dopamine   Release of Dopamine on Amphetamines

Amphetamines increase the amount of dopamine in the synapse, causing the cell to constantly fire.

....

2. *The drugs do this by both increasing the amount of dopamine released into the synapse and inhibiting the removal, or uptake, of dopamine from the synapse, making the chemical linger longer.*

cumbens and the ventral striatum. These cortical parts make up the dopamine reward pathway, the neural highway that's responsible for generating the pleasurable emotions triggered by pleasurable things. It doesn't matter if it's having sex or eating ice cream or snorting cocaine: these things fill us with bliss because they tickle these cells. Happiness begins here.

What's the purpose of all this pleasure? It turns out that the real benefit of delight—and the reason amphetamines increase creative production—is its powerful effect on *attention*. The same neurons that generate happiness (and get titillated by Benzedrine) also play a crucial role in determining which thoughts enter conscious awareness. A sense of pleasure is the brain's way of telling itself to look over there, or there, or there. The result is that dopamine acts like a neural currency—a price tag for information—allowing us to quickly appraise the outside world. The chemical tells us what we should notice, which things and thoughts are worth the cost of awareness.

The wiring of the brain reflects this evolutionary innovation: there's a highway of nerves connecting the pleasure center—the dopamine reward pathway—to the prefrontal cortex, a mass of tissue behind the forehead that controls attention. This is the area that allows someone to zoom in on reality, so that all he is thinking about is a single line in a single poem. Instead of getting distracted by the wandering mind, he can concentrate on the work. This essential mental talent depends on the prefrontal cortex and the squirts of dopamine that help guide one's gaze.[3]

....

3. *This also explains why Ritalin and other amphetamines are used to treat attention deficit hyperactivity disorder (ADHD). Because they make classroom tasks more rewarding, they make it easier for children to pay attention.*

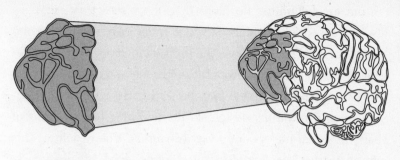

Prefrontal Cortex   *The prefrontal cortex helps direct the spotlight of attention, keeping us focused on the task at hand.*

And this is why amphetamines can be so helpful, at least for distractible writers. The drugs are a chemical shortcut. Because those dopamine neurons in the midbrain are so excited—they are suffused with the neurotransmitter—the world is suddenly saturated with intensely interesting ideas.[4] While attention is normally impatient and twitchy, flitting about from sensation to sensation, the drug-induced flood of dopamine makes even the most tedious details too interesting to ignore.[5] For Auden, the drug was a crucial tool, since all of his writing required endless revision. A single line was changed, and then changed again. Syllables were trimmed, rhymes altered, words cut.

....

4. *Daniel Kahneman and Jackson Beatty have demonstrated that any task requiring extended bouts of focused attention, such as editing a poem or solving an algebra problem, causes the pupils to dilate. "Much like the electricity meter outside your house or apartment, the pupils offer an index of the current rate which mental energy is used in the brain," Kahneman writes. It shouldn't be too surprising, then, that amphetamines also cause pupils to dilate, as the drug dramatically increases the amount of attention we can devote to the world.*
5. *Some people get a similar boost naturally; studies have linked small coding differences in the genes that underlie dopamine production, such as the COMT Val/Met polymorphism, to variations in attentional abilities, with more neurotransmitter equaling more attention.*

But amphetamines do more than focus the attention. They also make it easier to *connect* ideas, to translate concentration into better poetry. That's because the prefrontal cortex—that area in charge of attention—is also a theater of ideas, a mental space to store all of our pleasurable and interesting thoughts. As the human brain evolved, its design slowly morphing over millions of years, this area underwent a vast expansion. The benefit of this new anatomy was an unprecedented cognitive talent called *working memory*. The name is accurate: by keeping information in short-term storage, where it can be consciously contemplated, the prefrontal cortex allows us to work with all the fleeting thoughts flowing in from the various parts of the brain.

It's impossible to overstate the importance of working memory. For one thing, there is a strong correlation between working memory and general intelligence, with variations in the size of working memory accounting for approximately 60 percent of the variation in IQ scores. (Being able to cram more ideas into the prefrontal cortex makes a person smarter.) Thanks to working memory, we can play with abstractions and analyze poems; we can obsess over math problems and treat ideas like tangible things. Instead of just experiencing those rewarding sensations, we can *think* about them. The pleasure can be contemplated.

The power of working memory also explains why amphetamines are abused by poets and mathematicians seeking a creative edge. When we're intensely focused on something, more information is sent to the prefrontal cortex; the stage of consciousness gets even more crowded. (If working memory is normally like a string quartet, these drugs turn it into a loud orchestra.) This excess of ideas allows the neurons to form connections that have never existed before, wiring themselves into novel networks. From the perspective of the brain, these new connections are merely old thoughts that occur at the exact same time.

But these new ideas are not epiphanies. The connections made in working memory don't feel mysterious, like an insight, or shock us with their sudden arrival. Instead, these creative thoughts tend to be minor and incremental—one can efficiently edit a poem but probably won't invent a new poetic form. These differences are a function of human wiring: while the right hemisphere excels at remote associations, the prefrontal cortex is tuned to detect the local connections between related ideas. We might miss the forest, but we can clearly see each tree.

This focused thought process, fueled by the ecstatic firing of prefrontal neurons, is necessary for solving certain kinds of creative problems. Just look at Auden's Benzedrine poetry. Each line is taut and concise; there are no wasted words. He has distilled his grand ideas down to their poetic minimum, their expressive essence. And so we get the final section of "In Memory of W. B. Yeats," which ends with a series of short rhyming verses, modeled on the clipped quatrains of Blake:

> *In the deserts of the heart*
> *Let the healing fountain start*
> *In the prison of his days*
> *Teach the free man how to praise.*

The poem is about the importance of poetry. Auden is arguing that the writer, by harnessing the power of language, gives us the power to "praise," to illuminate a world shrouded in indifference. For Auden, Benzedrine was merely a means of perfecting the praise, a chemical tool that helped ensure he got the words exactly right.

To understand how such focused creativity differs from moments of insight, compare this spare stanza to the loose lyricism of Dylan's "Like a Rolling Stone." While Auden's rhymes are tight and direct—his elegy to Yeats includes *heart / start* and *dark / bark*—Dylan's syllabic associations are much more unexpected. (In one fevered stretch, he rhymes *compromise, alibis, realize,* and *eyes.*) And then there is the poetic content: Auden's lyrics unfold like a linear argument. Dylan, however, is channeling the expansive flow of his right hemisphere, "vomiting" forth lyrics that make little literal sense.

Friedrich Nietzsche, in *The Birth of Tragedy,* distinguished between two archetypes of creativity, both borrowed from Greek mythology. There was the Dionysian drive—Dionysus was the god of wine and intoxication—which led people to embrace their unconscious and create radically new forms of art. (As Dylan once said, "I accept the chaos. I hope it accepts me.") The Apollonian artist, by contrast, attempted to resolve the messiness and impose a sober order onto the disorder of reality. Like Auden, creators in the spirit of Apollo distrust the rumors of the right hemisphere. Instead, they insist on paying careful attention, scrutinizing their thoughts until they make sense. Auden put it best: "All genuine poetry is in a sense the formation of private spheres out of public chaos."

Modern science has given Nietzsche's categories a new set of names. The Dionysian innovator, trusting all those spontaneous epiphanies, is a perfect example of *divergent* thinking. He needs these unexpected thoughts when logic won't help, when working memory has hit the wall. In such instances, the right hemisphere helps expand the internal search. This is the kind of thinking that's essential when struggling with a remote associate problem, or trying to invent a new kind of pop song, or figuring out what to do with a weak glue. It's the thought process of warm showers and blue rooms, paradigm shifts and radical restructurings.

The Apollonian artist, by comparison, relies on *convergent* thinking. This mode of thought is all about analysis and attention. It's the ideal approach when trying to refine a poem, or solve an algebra equation, or perfect a symphony. In these instances, we don't want lots of stray associations — such thoughts are errant distractions. Instead, we want to focus on the necessary information, filling our minds with relevant thoughts. And so we slowly converge on the ideal answer, chiseling away at our errors. This process is a struggle, a long labor of attention, but that's the point. It takes time to find the perfect line.

1.

Earl Miller has devoted his career to understanding the prefrontal cortex, that warehouse of working memory. He has a shiny shaved head and a silver goatee. His corner office in the gleaming Picower Institute at MIT is cantilevered over an old freight-train track, so every afternoon the quiet hum of the lab is interrupted by the rattle of a locomotive. Miller's favorite word is *exactly* — it's the adverb that modifies everything, so a hypothesis was "exactly right" or an experiment was "exactly done" — and that emphasis on precision has defined his career. His first major scientific advance was a byproduct of necessity. It was 1995 and Miller had just started his lab, which meant that he had no money. His research involved recording the activity of neurons in the monkey brain, monitoring the flux of voltage within an individual cell as the animal performed various tasks. "There were machines that allowed you to record from eight or nine [neurons] at the same time, but they were very expensive," Miller says. "I still had no grants, and there was no way I could afford one." So Miller began inventing his own apparatus in his spare time. After a few months of tinkering, he constructed a messy tangle of wires, glass pipettes, and electrodes that could record simultaneously from numerous cells distributed across the monkey cortex. "It worked even better than the expensive machine," Miller says. "And then we just made the units smaller and smaller, which meant we could record from more and more neurons."

This methodological advance—it's known as multiple-electrode recording—allowed Miller to watch information zip around the brain as the electrical cells interacted with one another. Miller was most interested in studying the prefrontal cortex, though, since this brain area is such an aggregator of information. "It's where everything projects to," Miller says. "It's literally where the world comes together."

Because Miller can eavesdrop on neurons, he's been able to describe this flood of ideas at the most fundamental level. He can listen as cells in the prefrontal cortex struggle to make sense of the information in working memory, searching for relevant patterns and new connections. In one of Miller's current experiments, he shows monkeys a picture of randomly scattered dots, which look like stars in the night sky. This picture is the prototype. Then Miller flashes the monkeys a set of distorted pictures, in which the dots have been haphazardly shifted around the screen. The monkeys are required to indicate, using a joystick, whether or not these different pictures are similar in any way to the initial picture. At first, the monkeys guess randomly, and they learn from trial and error. They struggle to figure out the essence of the prototype and how to characterize it. Is the picture defined by the square of dots in the center? Or the cluster of dots off to the left? Interestingly, it turns out that even when the original picture is taken away—the monkeys can no longer see the prototype—the prefrontal cells devoted to that picture continue to fire. They're still holding on to that particular arrangement of dots, which is why working memory is a type of *memory*. (Scientists refer to this as RAM-like activity, since these brain cells are acting just like random-access memory in a computer.) This echo of activity lasts for only a little while, but it's long enough to mix together thoughts, as seemingly unrelated ideas intersect. And so, after a few minutes of staring at different pictures of dots, the monkeys are able to sort the pictures into categories and determine which ones most resemble the prototype. "At a certain point, the monkeys just get it," Miller says. "They suddenly realize that there are patterns here." This new idea—it's essentially an abstract rule for connecting the dots—is represented as a new circuit of neural activity in the prefrontal cortex. The brain cells have been literally altered by the breakthrough, changed by the creative connection.

While the prefrontal cortex is the source of these creative ideas, it doesn't generate these new connections by itself. Miller has discovered that instead, the prefrontal cortex works in close collaboration with other brain areas, such as the basal ganglia and dopamine reward pathway. The process goes like this: rewarding information—and the reward can be anything from a sweet treat to a poetic metaphor—gets processed by the dopamine neurons and then sent onward to the prefrontal cortex. The thought has now entered working memory. If this new information leads to any useful conclusions—if it allows the monkey to decipher the dots or helps a poet improve a poem—then the idea survives, a persistent link between cells. A new connection that helps solve a problem has been created.

But the process isn't finished. That new thought is then transmitted back to its source—those pleasure-hungry dopamine cells in the midbrain—so the neurons learn from the new idea. "We call that a recursive loop," Miller says. "It allows the system to feed on itself, so that one idea leads naturally to the next. We can then build on these connections, so that they lead to other, richer connections."[6]

This loop of creativity illuminates the power of attention. When each of us focuses on something, the idea enters working memory. As a result, we're able to slowly chisel away at our creative tasks. Perhaps it's finally finding the perfect choreography for a dance or figuring out how to solve the architectural problem. These unprecedented thoughts are then transmitted back down the line, so that the brain modifies its own sense of what's important. We suddenly look at reality through a slightly different lens, as the new idea is seamlessly incorporated into our perceptions. Instead of just seeing a scattering of dots, we notice the pattern; things are starting to make sense. We have stared at the world, and the world itself has changed.

2.

When Milton Glaser was sixteen, he decided to draw a portrait of his mother. "I was just sitting in front of her one night and I thought it would be fun to sketch her face," he says. "So I got out a piece of paper and a charcoal pencil. And you know what I realized? I realized I hadn't the faintest idea what she looked like. Her image had become fixed in my mind at the age of one or two, and

6. One of the interesting implications of Miller's work is the importance of the primitive midbrain in the creative process. While this brain area is often disparaged—it's seen as the primal source of rewards, not the center of epiphanies—Miller's research demonstrates that the midbrain also plays a crucial role in helping to locate the relevant information that will help solve a problem. "The basal ganglia and these other areas are the engine behind so many higher cognitive functions," Miller says. "They may be more primitive, but they are what grasp the pieces of the puzzle. Only then can the pieces be sent along to the prefrontal cortex, which puts the puzzle together."

it really hadn't changed since. I was drawing a picture of a woman who no longer existed."

But as Glaser stared at her face and then compared what he saw to the black marks on the paper, her appearance slowly came into view. He was able to draw her as she was, and not as he expected her to be. "That sketch taught me something interesting about the mind," he says. "We're always looking, but we never really see." Although Glaser had looked at his mother every single day of his life, he didn't *see* her until he tried to draw her. "When you draw an object, the mind becomes deeply, intensely attentive," Glaser says. "And it's that act of attention that allows you to really grasp something, to become fully conscious of it. That's what I learned from my mother's face, that drawing is really a kind of thinking."

Milton Glaser looks like a patriarch from a Philip Roth novel. His bare head is ringed with gray hair; oversize glasses rest on the long slope of his nose. Glaser is eighty years old, but he still works in a small studio on East Thirty-second Street in Manhattan. It's a cluttered space, the white walls hidden by old art posters, colorful prints for 1980s rock concerts, and art books stacked ten high. Above the front door, chiseled into the glass, is the slogan of the studio: ART IS WORK.

For Glaser, the quote summarizes his creative philosophy. "There's no such thing as a creative *type*," he says. "As if creative people can just show up and make stuff up. As if it were that easy. I think people need to be reminded that creativity is a *verb*, a very time-consuming verb. It's about taking an idea in your head, and transforming that idea into something real. And that's always going to be a long and difficult process. If you're doing it right, it's going to feel like work."

Glaser is a living legend in the world of graphic design, having created a number of the most iconic illustrations of the twentieth century, from the I ♥ NY ad campaign to the 1967 Bob Dylan silhouette poster. He came up with the DC Comics logo, cofounded *New York* magazine, and invented numerous typefaces; he's designed the interiors of famous restaurants and is responsible for a staggering number of product labels on the supermarket shelves. In recent years, his images have entered the permanent collections of MOMA, the Smithsonian, and the Cooper-Hewitt National Design Museum.

But Glaser almost didn't make it. In the late 1950s, when he began working on Madison Avenue, photography seemed like the ad form of the future. Although print ads had once relied on trained illustrators — they helped spin the fantasy — the industry was transitioning to staged photo ops. "People like me seemed so antiquated," Glaser remembers. "Why draw something when you could just take a picture of it, or make a television commercial?" The camera was king; artists were out of work.

As a cofounder of Pushpin Studios, however, Glaser helped rediscover the potential of graphic design. He introduced bright neon colors and a touch of abstraction; he found a way to turn even staid commissions into conceptual works of art. Glaser knew that the most powerful images weren't the most realistic. Instead of simply trying to represent a thing, Glaser wanted to *define* it. His perfect visual was more than a picture: it was a summary of associations, a map of thought. It was a picture honed by human attention.

The creative possibilities of graphic art are perfectly captured by Glaser's most iconic design. In 1975, he accepted an intimidating assignment: create an ad campaign that would rehabilitate the image of New York City. At the time, Manhattan was falling apart. Crime was at an all-time high, and the city was almost bankrupt.

"When people thought about the city, they thought about dirt and danger," Glaser remembers. "And they wanted a little ad campaign that could somehow change all that." There was one additional constraint: the print ad had to use the phrase *I love New York*.

Glaser began by experimenting with fonts, laying out the straightforward slogan in a variety of friendly typefaces. After a few weeks of work, he settled on a charming cursive, with *I Love NY* set against a plain white background. "I send in my proposal and it's approved," Glaser says. "Everybody likes it. And if I were a normal person, I'd stop thinking about the project. But I can't. Something about it just doesn't feel right."

So Glaser continues to fixate on the design, devoting hours to a project that was supposedly finished. "I can't get the damn problem out of my head," he says. "And then, about a week after the first concept was approved, I'm sitting in a cab, stuck in traffic. I often carry spare pieces of paper in my pocket, and so I get the paper out and I start to draw. And I'm thinking and drawing and then I get it. I see the whole design in my head. I see the typewriter typeface and the big round red heart smack-dab in the middle. I know that this is how it should go."

For Glaser, the I ♥ NY ad is a testament to the importance of persistence. Because he refused to stop thinking about the three-word slogan—he kept on redrawing the logo in his mind—his ideas continued to improve. And then, while stuck in the taxi, this steadfast focus led to a new design, a better design. The graphic that he imagined in rush-hour traffic has become the most widely imitated work of graphic art in the world.

This is the power of attention and working memory; it allows us to relentlessly refine our ideas, to continue thinking about our thoughts. "Design is the conscious imposition of meaningful order," Glaser says. "That sounds grandiose, but it's just the process of taking an idea that isn't clear and making it a little more clear. I could tell you a bullshit story about what exactly led to the idea [of I ♥ NY], but the truth is that I don't know. Maybe I saw a red heart out of the corner of my eye? Maybe I heard the word? But that's the way it always works. You keep on trying to fix it, to make the design a little bit more interesting, a little bit better. And then, if you're really stubborn and persistent and lucky, you eventually get there."

Glaser's impressive work ethic—his ability to stick with a problem until it surrenders—is itself a skill that took years to develop. In 1951, Glaser was an impressionable twenty-one-year-old with a Fulbright fellowship who was heading to Bologna to study etching with the painter Giorgio Morandi. At the time, Morandi was creating his *natura morta* paintings, a collection of still lifes that featured empty wine bottles and terra-cotta vases set against a flat gray background. The art was austere, a reflection of Morandi's disciplined artistic process. He spent months on each canvas, trying to edge closer to the fragile reality he wanted to describe. Sometimes, Morandi would just stare at that random collection of containers and become too intimidated to paint. "I'd watch him get so focused on these incredibly tiny details," Glaser remembers. "He'd devote weeks of his life to moving a passage of gray a quarter of an inch to the left, or smoothing out the curve of a bottle. It didn't matter that nobody else would notice. He would notice, and that was more than enough."

Morandi's obsessive dedication to getting the image exactly right changed forever the way Glaser thought about creativity. His old artistic model had been Pablo Picasso—"A raging lunatic genius who wanted to devour the world," as Glaser puts it—but his new hero was the modest Italian painter. "It was Morandi who taught me about dedication," Glaser says. "He showed me the necessity of persistence, and that nothing good is ever easy. And that's because we see nothing at first glance. It's only by really thinking about something that we're able to move ourselves into perceptions that we never knew we had the capacity for."

The German philosopher Martin Heidegger referred to this as the unconcealing process. He argued, like Glaser, that the reality of things is naturally obscured by the clutter of the world, by all those ideas and sensations that distract the mind. The only way to see through this clutter is to rely on the knife of conscious attention, which can cut away the excess and reveal "the things themselves."

This mental process—the act of unconcealing—has been an essential tool for Glaser. He relies on the solvent of working memory to turn difficult assignments into resonant images that precisely convey the necessary set of associations. This helps explain why Glaser has no defining style. He isn't a minimalist or a maximalist, a realist or a cartoonist. ("I distrust styles," he says. "To have a style is to be trapped.") Instead, Glaser treats each assignment as a unique problem with its own requirements and constraints. He doesn't know what the answer will look like or what kind of image it will require. That's why he needs to think about it.

Consider the Brooklyn Brewery project. In 1987, Milton Glaser was approached by Steve Hindy and Tom Potter, two businessmen and beer aficionados interested in reviving the great tradition of Brooklyn brewing. They asked Glaser to design a logo for their new company that would somehow capture the spirit of the borough. "Their first idea was to use the Brooklyn Bridge or to call the beer the Brooklyn Eagle," Glaser remembers. "And that could have worked; that's clever enough. But I told them, 'Why settle for only a small piece of Brooklyn when you can own the whole place?'"

And so Glaser meditated for days on Brooklyn and craft beer. The first part of the solution arrived when Glaser misremembered the baseball cap of the Brooklyn Dodgers. He drew a wide, arcing *B* that he thought was the team's logo. (The Dodgers actually used a staid, upright font; Glaser had embellished the letter in his memory.) "It didn't matter that the logo wasn't accurate," he says. "It *felt* like it was from baseball. It really made you think of the Dodgers."

But Glaser knew that something was missing; the logo was incomplete. "And so I'm racking my brain, because I've got to give the design a European feel," he says. "It's got to feel like an import beer with a distinguished history, not another Budweiser." After weeks of market research, analyzing all the beer bottles he could find, Glaser settled on a simple frame of white dots on a green background with the big *B* in the center. "I can't tell you why that logo works," he says. "I can't explain how it came to me. But I just kept on playing with the design until I found what I needed."[7]

This is Glaser's fundamental method: he thinks until he can think no more, until his attention gives out or his solution is unconcealed. Of course, such a process is possible only because those cells in the prefrontal cortex can cling to ideas, allowing one

----

7. When Glaser first showed the founders of the Brooklyn Brewery his proposed logo, they were mostly disappointed. "The greatly anticipated logo was a shock," remembers Steve Hindy. "There was no Brooklyn Bridge. There was no soaring eagle. It was just a B." Glaser told the businessmen to give the image a chance, to "put it on the counter of the kitchen and live with it for a while." After a few days, says Hindy, the brilliance of the design began to sink in. "The B evoked the nostalgia of the Dodgers while being a fresh symbol of Brooklyn . . . It looked like the logo of a company that had been in business for decades."

to focus on abstractions like beer logos and the meaning of Brooklyn. The logo itself is a visual hybrid, a literal overlapping of ideas. There is the cursive *B* that makes us think of baseball, but there is also the cosmopolitan background echoing the labels of old European beers. Once these distinct visuals collided in his working memory, Glaser knew that his problem was solved: he'd found a pretty image that conjured up the ideas he needed to convey. And it fit on a beer bottle.

3.

The lesson of W. H. Auden and Milton Glaser is that working memory is an essential tool of the imagination. Sometimes, all we need to do is pay attention, to think until the necessary thoughts intersect. The progress will be slow, but the answer will gradually reveal itself, like a poem emerging from the edits. As Nietzsche observed in his 1878 book *Human, All Too Human:*

> Artists have a vested interest in our believing in the flash of revelation, the so-called inspiration . . . shining down from heavens as a ray of grace. In reality, the imagination of the good artist or thinker produces continuously good, mediocre, or bad things, but his judgment, trained and sharpened to a fine point, rejects, selects, connects . . . All great artists and thinkers are great workers, indefatigable not only in inventing, but also in rejecting, sifting, transforming, ordering.

To illustrate his point, Nietzsche described Beethoven's musical notebooks, which documented the composer's painstaking process of refining his melodies. It wasn't uncommon for Beethoven to experiment with seventy different versions of a phrase before settling on the final one. "I make many changes, and reject and try again, until I am satisfied," the composer re-marked to a friend. In other words, even Beethoven—the cliché of artistic genius—needed to constantly refine his ideas, to strug-gle with his music until the beauty shone through.

Although this mental talent is an essential part of the creative process, there is nothing easy or pleasant about it. It isn't fun re-fining a musical motif, or cutting a favorite line, or throwing away a sketch. In fact, there is evidence suggesting that the ability to relentlessly focus on a creative problem can actually make us mis-erable. Aristotle was there first, stating in the fourth century B.C. that "all men who have attained excellence in philosophy, in po-etry, in art and in politics, even Socrates and Plato, had a melan-cholic habitus; indeed some suffered even from melancholic dis-ease." This belief was revived during the Renaissance, leading Milton to exclaim, in his poem "Il Penseroso": "Hail divinest mel-ancholy / whose saintly visage is too bright / to hit the sense of hu-man sight." The Romantic poets took the veneration of sadness to its logical extreme and described suffering as a prerequisite for the literary life. As Keats wrote, "Do you not see how necessary a world of pains and troubles is to school an intelligence and make it a soul?"

Joe Forgas, a social psychologist at the University of New South Wales, in Australia, has spent the last decade investigating the link between negative moods and creativity. Although people tend to disparage sadness and similar moods, Forgas has repeat-edly demonstrated that a little melancholy sharpens the spotlight of attention, allowing us to become more observant and persis-tent. (Of course, feeling sad also makes us less likely to have mo-ments of insight.) His most compelling study took place in a sta-tionery store in the suburbs of Sydney. The experiment itself was simple: Forgas placed a variety of trinkets such as toy soldiers, plastic animals, and miniature cars near the checkout counter. As shoppers exited, Forgas tested their memory, asking them to list as many of the items as possible. To control the mood of the subjects, when Forgas conducted the survey on gray, rainy days, he accentuated the weather by playing Verdi's Requiem; on sunny days, he used a chipper soundtrack of Gilbert and Sullivan. The results were clear: shoppers in the "low mood" condition remembered nearly four times as many of the trinkets. The wet weather made them sad, and their sadness made them more attentive.

These are the same cognitive skills that underlie the unconcealing process: the negative mood acts like a mild dose of amphetamine. When someone is immersed in melancholy, it's easier for him to linger on a poetic line or keep thinking about a beer logo. This helps explain why Forgas has found that states of sadness—he induces the downcast mood with a film about death and cancer—also correlate with better writing samples; subjects compose sentences that are clearer and more compelling. Because they were more attentive to what they were writing, they produced more refined prose, the words polished by their misery.

Modupe Akinola, a professor at Columbia Business School, has expanded on these provocative results. In one of her most recent experiments, she asked each subject to give a short speech about his or her dream job. The students were randomly assigned to either a positive- or negative-feedback condition; in the positive-feedback condition, speeches were greeted with smiles and vertical nods, and in the negative, speeches met frowns and horizontal shakes. After the speech was over, the subject was given glue, paper, and colored felt and told to make a collage using the materials. Professional artists then evaluated each collage according to various metrics of creativity.

Not surprisingly, the feedback affected the mood of the subjects: those who received smiles during their speeches reported feeling better than before, while frowns had the opposite effect. What's interesting is what happened next. Subjects in the negative-feedback condition created much prettier collages. Their angst led to better art. As Akinola notes, this is largely because the sadness improved their focus and made them more likely to persist with the creative challenge. As a result, they kept on rearranging the felt, playing with the colorful designs.

The enhancement of these mental skills during states of sadness might also explain the striking correlation between creativity and depressive disorders. In the early 1980s, Nancy Andreasen, a neuroscientist at the University of Iowa, interviewed several dozen writers from the Iowa Writers' Workshop about their mental history. While Andreasen expected the artists to suffer from schizophrenia at a higher rate than normal—"There is that lingering cliché about madness and genius going together," she says—that hypothesis turned out to be completely wrong. Instead, Andreasen found that 80 percent of the writers met the formal diagnostic criteria for some type of depression. These successful artists weren't crazy—they were just exceedingly sad. A similar theme emerged from biographical studies of British novelists and poets done by Kay Redfield Jamison, a professor of psychiatry at Johns Hopkins. According to her data, famous writers were eight times as likely as people in the general population to suffer from major depressive illness.

Why is severe sadness so closely associated with creativity? Andreasen argues that depression is intertwined with a "cognitive style" that makes people more likely to produce successful works of art. Her explanation is straightforward: It's not easy to write a good novel or compose a piece of music. The process often requires years of careful attention as the artist fixes mistakes and corrects errors. As a result, the ability to stick with the process—to endure the unconcealing—is extremely important. "Successful writers are like prizefighters who keep on getting hit but won't go down," Andreasen says. "They'll stick with it until it's right. And that seems to be what the mood disorders help with." While Andreasen acknowledges the terrible burden of mental illness—she quotes Robert Lowell on depression not being a "gift of the Muse" and describes his reliance on lithium to escape the pain—she argues that, at least in its milder forms, the disorder benefits many artists due to the perseverance it makes possible.[8] "Unfortunately, this type of thinking is often inseparable from the suffering," Andreasen says. "If you're at the cutting edge, then you're going to bleed."

While the unconcealing process allows us to refine our creations, not every problem benefits from melancholy, amphetamines, and the prefrontal cortex. Milton Glaser has learned that his hardest assignments often require a flash of insight before he can begin lavishing the project with attention. "You want to make sure you're focused on the right question," Glaser says. "Sometimes that means you've got to begin the process by trusting an intuition that you can't explain."

The necessary interplay of these different creative modes—the elation of the insight, and the melancholy of the unconcealing—begins to explain why bipolar disorder, an illness in which people oscillate between intense sadness and extreme euphoria, is so closely associated with creativity. Andreasen found that nearly 40 percent of the successful creative people she investigated had

---

8. *Severe depression, however, is clearly useless—the pain is so intense that nothing can be done.*

the disorder, a rate that's approximately *twenty times* higher than it is in the general population. (More recently, the psychiatrist Hagop Akiskal found that nearly two-thirds of a sample of influential European artists were bipolar.) The reason for this correlation, Andreasen suggests, is that the manic states lead people to erupt with new ideas as their brains combust with remote associations. "When people are manic, they are driven by this intense, overwhelming need to express themselves," Andreasen says. "It can be an extremely unpleasant condition, often because they can't stop creating." Furthermore, these imaginative outpourings are defined by their radical, sometimes incomprehensible, nature. "People become much more open to unexpected ideas when manic," Andreasen notes. "That's when they typically come up with their most original concepts."

And then the mania ebbs. The extravagant high descends into a profound low. While this volatility is horribly painful, it can also enable creativity, since the exuberant ideas of the manic period are refined during the depression. In other words, the emotional extremes of the illness reflect the extremes of the creative process: there is the ecstatic generation phase, full of divergent thoughts, and the attentive editing phase, in which all those ideas are made to converge.[9] This doesn't take away, of course, from the agony of the mental illness, and it doesn't mean that people can create only when they're horribly sad or manic. But it does begin to explain the significant correlations that have been repeatedly observed between depressive syndromes and artistic achievement.

The larger lesson is that different kinds of creative problems benefit from different kinds of creative thinking. (T. S. Eliot un-

....

9. *This has led several psychiatrists, such as Andy Thomson at the University of Virginia, to speculate that the enhanced creativity of people with bipolar disorder might explain why the illness has been preserved throughout human evolution.*

derstood this: "The bad poet is usually unconscious where he ought to be conscious, and conscious where he ought to be unconscious.") The question, of course, is how to adjust our thought process to the task at hand. How does anyone know when to listen to the prefrontal cortex instead of unleashing the right hemisphere? When is it time to daydream and take warm showers, and when is it better to drink another cup of coffee?

The good news is that the human mind has a natural ability to diagnose its own problems, to assess the kind of creativity that's needed. These assessments have an eloquent name: they're called "feelings of knowing," and they occur when we suspect that we can find the answer if only we keep on thinking about the question. Consider tip-of-the-tongue moments. It's estimated that people experience these, on average, about once a week.[10] Perhaps it occurs when you run into an old acquaintance whose name you can't remember although you know what letter it begins with. Or perhaps you fail to recall the title of a recent movie, even though you can describe the plot in perfect detail. What's interesting about this cognitive hiccup is that, although you can't remember the information, you're still convinced that you know it, which is why you devote so much mental energy to trying to recover the missing word.

But here's the mystery: If you've forgotten a person's name, then why are you so sure that you can remember it? What does it mean to know something without being able to access it? This is where feelings of knowing prove essential. Numerous studies have demonstrated that when it comes to problems that don't require insights, the mind is remarkably accurate at assessing the

---

10. *This is universal, and the vast majority of languages, from Afrikaans to Hindi to Arabic, even rely on tongue metaphors to describe the tip-of-the-tongue moment.*

likelihood that a problem can be solved. You can glance at a question and know that the answer is within reach if only you put in the work. As a result, you're motivated to stay focused on the challenge.

What makes these feelings of knowing even more useful is that they come attached to a sense of progress. This was first demonstrated by Janet Metcalfe, a psychologist at Columbia University. She asked people working on various creative puzzles whether or not they felt like they were getting closer ("warmer") to the solution. When the subjects were working on problems that were typically solved with insights, they reported no increase in warmth until the insights popped into their heads—they went straight from cold to burning hot. There was no feeling of knowing. In contrast, Metcalfe found, problems that didn't require insight were answered only after people reported a gradual increase in warmth, which reflected their sense of progress. What's impressive about such estimates is that people were able to assess their closeness to the solution without knowing what the solution was.

This ability to calculate progress is an important part of the creative process. When you don't feel that you're getting closer to the answer—you've hit the wall, so to speak—you probably need an insight. In these instances, you should rely on the right hemisphere, which excels at revealing those remote associations. Continuing to focus on the problem will be a waste of mental resources, a squandering of the prefrontal cortex. You will stare at your computer screen and repeat your failures. Instead, find a way to relax and increase the alpha waves. The most productive thing to do is forget about work.

However, when those feelings of knowing tell you that you're getting closer—when you feel the poetic meter slowly improve, or sense that the graphic design is being unconcealed—then you need to keep on struggling. Continue to pay attention until it hurts; fill your working memory with problems. Before long, that feeling of knowing will become actual knowledge.

There is nothing romantic about this kind of creativity, which consists mostly of sweat, sadness, and failure. It's the red pen on the page and the discarded sketch, the trashed prototype and the failed first draft. It's ruminating in the backs of taxis and popping pills until the poem is finished. Nevertheless, such a merciless process is sometimes the only way forward. And so we keep on thinking, because the next thought might be the answer.

4 THE LETTING GO

*The struggle of maturity is to recover the
seriousness of a child at play.*
—Friedrich Nietzsche

THE THEATER IS EMPTY; the house lights are low. Yo-Yo Ma is
lugging his cello across the stage toward a lonely metal chair at its
center. The instrument looks heavy, and Ma takes delicate steps,
the long horsehair bow jutting out into space. He sits, steadies
himself in the chair, and stares for a long moment at the sheet
music. Then he raises his right arm, positions his fingers on the
wooden neck, and drags the bow across the strings. The first note
sounds like a beautiful moan.

I'm sitting next to Bruce Adolphe, the composer of the piece
Ma is rehearsing, and he seems a little nervous. Because Ma is
such a celebrity—in the previous two months, he's played twenty-
three concerts in eighteen cities—this is the first time Adolphe
has heard him play the music. "There's always that anxiety that
comes during the run-throughs," Adolphe says. "I've been living
with these notes for so long, but it always sounds different when
it's up on stage." Ma is sight-reading the piece, so he begins play-

ing slowly, like someone trying to decipher the first pages of a novel written in a barely familiar language. Sometimes he stops in the middle of a phrase and then repeats the notes with a slightly different interpretation.

And then, after a few tentative minutes, Ma begins to disappear into the music. I see it first in his body, which begins to subtly sway. The movement then spreads to his right arm, so that the bow starts to trace wider and wider arcs in the air. Before long, Ma's shoulders are relaxed and expressive, drawing together whenever the tempo increases. And when he repeats the theme of the piece, his eyes briefly close, as if he were entranced by the same beauty he's pouring into space. I look over at Adolphe: his tension has turned into a faint smile.

Bruce Adolphe first met Ma at the Juilliard School in New York City. Although Ma was only fifteen years old at the time, he was already an established performer, having played for JFK at the White House and with Leonard Bernstein on national television. Adolphe was a promising young composer who had just written his first cello piece. "Unfortunately, I had no idea what I was doing," Adolphe remembers. "I'd never written for the instrument before." He'd shown a draft of his composition to a Juilliard instructor, who told him that the piece featured a chord that was impossible to play. Before Adolphe could correct the music, however, Ma decided to rehearse the composition in his dorm room. "Yo-Yo played through my piece, sight-reading the whole thing," Adolphe says. "And when that impossible chord came, he somehow found a way to play it. His bow was straight across all four strings. Afterward, I asked him how he did it, because I had been told by the teacher that it couldn't be done. And Yo-Yo said,

'You're right. I don't think it can be done.' And so we started over again, and this time when the chord came I yelled, 'Stop!' We both looked at his left hand, and it was completely contorted on the fingerboard. The hand position he had somehow found was uncomfortable for him to hold; his fingers were twisted in a most unnatural way. 'See,' Yo-Yo said, 'you're right, you really can't play that.' But he did!"

For Adolphe, the story is a reminder of Ma's astonishing talent, his ability to play those unplayable chords. It's a virtuosity that has turned Ma into one of the most famous classical performers in the world, an artist celebrated for a wide variety of recordings, from the cello suites of Bach to the swing of American bluegrass. He's improvised with Bobby McFerrin, recorded scores for Hollywood blockbusters, and popularized the melodies of Central Asia. "Sometimes, I'll watch him play and I'll feel that same awe I felt as a student at Juilliard," Adolphe says. "He can take your notes and he can find the thing that makes them come alive. Ma is a technical master, of course, but what makes him such a special performer is that he also knows when to release technique for something deeper, for that depth of emotion that no one else can find."

But Ma wasn't always such an expressive performer. In fact, his pursuit of musical emotion began only after a memorable failure. "I was nineteen and I had worked my butt off," Ma told David Blum of *The New Yorker* in 1989. "I knew the music inside and out. While sitting there at the concert, playing all the notes correctly, I started to wonder, 'Why am I here? What's at stake? *Nothing.* Not only is the audience bored but I myself am bored.' Perfection is not very communicative." For Ma, the tedium of the flawless performance taught him that there is often a tradeoff between perfection and expression. "If you are only worried about not making a mistake, then you will communicate nothing," he says. "You will have missed the point of making music, which is to make people feel something."

This search for emotion shapes the way Ma approaches every concert. He doesn't begin by analyzing the details of his cello part or by glancing at what the violins are supposed to play. Instead, he reviews the complete score, searching for the larger story. "I always look at a piece of music like a detective novel," Ma says. "Maybe the novel is about a murder. Well, who committed the murder? Why did he do it? My job is to retrace the story so that the audience feels the suspense. So that when the climax comes, they're right there with me, listening to my beautiful detective story. It's all about making people care about what happens next."

Ma's unusual musical approach is apparent during these rehearsals, as he carefully refines his interpretations of Adolphe's score. Over the course of the afternoon, his performance steadily accumulates its feeling; his body grows more loose-limbed and expressive. Ma's slight shifts of interpretation—hushing a pianissimo even more, speeding up a melodic riff, exaggerating a crescendo—turn a work of intricate tonal patterns into a passionate narrative. These shifts are not in the score, and yet they reveal what the score is trying to say. Most of the time, Ma can't explain what inspired these changes, but that doesn't matter: he has learned to trust himself, to follow his storytelling instincts.

And this is why Ma sways as he plays: Because he can't restrain himself. Because he is experiencing the same emotions that he is trying to express. Because he is letting himself go. "The best storytellers always get really into their own stories," Ma says. "They're waving their arms, laughing at their own jokes. That's what I try to be like on stage . . . I know that some of the best music happens when you let yourself get a little carried away."[1]

To make this kind of performance possible, Ma cultivates an easy, casual air backstage. Thirty minutes before the concert begins, Ma disappears into a quiet room. When he reemerges, I expect him to be somber and serious and maybe a little nervous. Instead, Ma is just as disarming and funny as ever, teasing me about my tie, eating a banana, and making small talk with Adolphe. This ease is not a pose: Ma needs to stay relaxed. If he is too clenched with focus, too edgy with nerves, then the range of his musical expression will vanish. He will not be able to listen to those feelings that guide his playing.

"People always ask me how I stay loose before a performance," Ma says. "The first thing I tell them is that everybody gets nervous. You can't help it. But what I do before I walk onstage is I pretend that I'm the host of a big dinner party, and everybody in the audience is in my living room. And one of the worst things you can do as a host is to show you're worried. Is the fish overcooked? Is the wine too warm? Is the beef too rare? If you show that you're worried, then everybody feels uncomfortable. This is what I learned from Julia Child. You know, she would drop her roast chicken on the floor, but did she scream? Did she cry or panic? No, she just calmly picked the chicken off the floor and managed to keep her smile. Playing the cello is the same way. I will make a mistake on stage. And you know what? *I welcome that first mistake.* Because

....

1. In many respects, Ma's obsession with spontaneity and expression—and his disinterest in perfection—evokes an earlier mode of performance. The classical music of the eighteenth century, for instance, is full of cadenzas, those brief parentheses in the score where the performer is supposed to play "freely." (The practice peaked with Mozart, who wrote cadenzas into most of his compositions.) In these frantic and somewhat unscripted moments, the performer was able to become a personality and express what he felt.

then I can shrug it off and keep smiling. Then I can get on with the performance and turn off that part of the mind that judges everything. I'm not thinking or worrying anymore. And it's when I'm least conscious of what I'm doing, when I'm just lost in the emotion of the music, that I'm performing at my best."

1.

There is something scary about letting ourselves go. It means that we will screw up, that we will relinquish the possibility of perfection. It means that we will say things we didn't mean to say and express feelings that we can't explain. It means that we will be onstage and not have complete control, that we won't know what we're going to play until we begin, until the bow is drawn across the strings.

While this spontaneous method might be frightening, it's also an extremely valuable source of creativity. In fact, the act of letting go has inspired some of the most famous works of modern culture, from John Coltrane's saxophone solos to Jackson Pollock's drip paintings. It's Miles Davis playing his trumpet on *Kind of Blue*—most of the album was recorded on the very first take—and Lenny Bruce inventing jokes at Carnegie Hall. Although this kind of creativity has always been defined by its secrecy, we are now beginning to understand how it happens.

The story begins in the brain. Charles Limb, a neuroscientist at Johns Hopkins University, has investigated the mental process underlying improvisation. Limb is a self-proclaimed music addict—he has a small recording studio near his office—and has long been obsessed with the fleshy substrate of creative performance. "How did Coltrane do it?" Limb asks. "How did he get up there onstage and improvise his music for an hour or sometimes more? Sure, a lot of musicians can throw out a creative little ditty here and there, but to continually produce masterpiece after masterpiece is nothing short of remarkable. I wanted to know how that happened."[2]

Although Limb's experiment was simple in concept—he was going to watch jazz pianists improvise new tunes while in a brain scanner—it proved difficult to execute. That's because the giant superconducting magnets in fMRI machines require absolute stillness of the body part being studied, which meant that Limb needed to design a custom keyboard that could be played while the pianists were lying down. (The setup involved an intricate system of angled mirrors, so the subjects could see their hands.) Each musician began by playing pieces that required no imagination, such as the C-major scale and a simple blues tune memorized in advance. But then came the creativity condition: the subject was told to improvise a new melody as she played along with a recorded jazz quartet.

While the subject was riffing on the keyboard, the scanner was monitoring minor shifts in brain activity. The scientists found that jazz improv relied on a carefully choreographed set of mental events. The process started with a surge of activity in the medial prefrontal cortex, an area at the front of the brain that is closely

....

2. *The birth of jazz improv is often traced back to Charles "Buddy" Bolden, an early-twentieth-century cornetist and one of the most popular musicians in New Orleans. In 1907, Bolden had a psychotic break, and he spent the rest of his life in a mental institution. (He was buried in an unmarked grave in 1930.) According to Dr. Sean Spence, a psychiatrist at the University of Sheffield, Bolden suffered from dementia praecox, an illness that was later classified as a variant of schizophrenia. Spence speculates that Bolden's unrelenting "madness"—he was hospitalized after threatening to attack people in the street—was actually a crucial inspiration for his "madcap" musical improv. His disordered thoughts, combined with his inability to read music, allowed him to arrange notes in a new way. Subsequent studies have found a disconcerting correlation between success in jazz and mental illness, from the heroin addiction of Charlie Parker to the erratic moods of Thelonious Monk to the debilitating depression and phantom-limb pain of Cole Porter.*

associated with self-expression. (Limb refers to it as the "center of autobiography" in the brain.) This suggests that the musician was engaged in a kind of storytelling, searching for the notes that reflected her personal style.

At the same time, the scientists observed, there was a dramatic shift in a nearby circuit, the dorsolateral prefrontal cortex (DLPFC). While the DLPFC has many talents, it's most closely associated with impulse control. This is the bit of neural matter that keeps each of us from making embarrassing confessions, or grabbing at food, or stealing from a store. In other words, it's a neural restraint system, a set of handcuffs that the mind uses on itself.

What does self-control have to do with creative improvisation? Before a single note was played in the improv condition, each of the pianists exhibited a "deactivation" of the DLPFC, as the brain instantly silenced the circuit. (In contrast, this area remained active when the pianist played a memorized tune.) The musicians were inhibiting their inhibitions, slipping off those mental handcuffs. According to Limb, this allowed them to create new music without worrying about what they were creating. They were letting themselves go.

But unleashing the mind is not enough — successful improv requires a very particular kind of creative expression. After it slips off the handcuffs, the brain must still find something interesting to say. This is the generation phase of the improv process, in which performers unleash a flood of raw material. What's so astonishing about this creative production, however, is that it's not reckless or random. Instead, the spontaneously generated ideas are constrained by the particular rules of the form. The jazz pianists, for instance, needed to improvise in the right key and tempo and mode. Jackson Pollock had to drip the paint in a precise pattern across the canvas. Or look at Yo-Yo Ma: his emotional release always fits the exacting requirements of the music. He sways, but he sways in perfect time. "I think the best way to perform is when your unconscious is fully available to you, but you're still a little conscious too," Ma says. "It's like when you're lying in bed in the early morning. I always have my best ideas then. And I think it's because I'm still half-asleep, listening to what my unconscious is telling me. But at the same time, I'm not in the midst of some crazy dream, because then it's just crazy. I guess it's a controlled kind of craziness. That's the ideal state for performance."

How does the brain find this liminal space? That was the question asked in a recent fMRI study by neuroscientists at Harvard in which twelve classically trained pianists were told to invent melodies. Unlike the Limb study, which compared brain activity during improv and memorized piano pieces, this experiment was designed to compare brain activity during different kinds of improv. (This would allow the scientists to detect the neural substrate shared by *every* form of spontaneous creativity, not just those bits of brain associated with particular types of music.) As expected, the various improv conditions—regardless of the musical genre—led to a surge of activity in a variety of neural areas, including the premotor cortex and the inferior frontal gyrus. The premotor activity is simply an echo of execution, as the new musical patterns are translated into bodily movements. The inferior frontal gyrus, however, is most closely associated with language and the production of speech. Why, then, is it so active when people compose on the spot? The scientists argue that expert musicians invent new melodies by relying on the same mental muscles used to create a sentence; every note is like a word. "Those bebop players play what sounds like seventy notes within a few seconds," says Aaron Berkowitz, the lead author on the Harvard study. "There's no time to think of each individual note. They have to have some patterns in their toolbox."

Of course, the development of these patterns requires years of practice, which is why Berkowitz compares improvisation to the learning of a second language. At first, he says, it's all about the vocabulary words; students must memorize a dizzying number of nouns, adjectives, and verb conjugations. Likewise, musicians need to immerse themselves in the art, internalizing the intricacies of Shostakovich or Coltrane or Hendrix. After musicians have studied for years, however, the process of articulation starts to become automatic — the language student doesn't need to contemplate her verb charts before speaking, just as the musician can play without worrying about the movement of his fingers. It's only at this point, *after* expertise has been achieved, that improvisation can take place. When the new music is needed, the notes are simply there, waiting to be expressed. It looks easy because they have already worked so hard.

These cortical machinations reveal the wonder of improvisation, the mirrors and wire behind this magic trick of creativity. They capture a mind able to selectively silence that which keeps us silent. And then, just when we've found the courage to create something new, the brain surprises us with a perfectly tuned burst of expression. This is what we sound like when nothing is holding us back.

2.

Clay Marzo has been waiting all morning for waves. He's standing with his surfboard next to a NO TRESPASSING sign on the edge of a pineapple field, looking down at a remote beach on the northwest shore of Maui. There are no tourists here because there is no sand, just a field of jagged lava rocks and a private dirt road. The tide is still too far out, so the waves are trashy; Clay hasn't said a word for more than an hour. He hasn't even moved. He's just been standing in the hot sun, staring at the sea.

The waiting ends shortly after 1:00 P.M., when the trade winds begin to blow. Clay rubs his hands together furiously, like a man trying to start a fire, and lets out a few guttural whoops. He then grabs his board and quickly descends the steep slope in his bare feet. There are a few surfers in the breaks to the right, away from the rocks. Clay heads to the left, where the waves are bigger. He paddles out and starts scanning the horizon, counting the seconds between the heaving swells. After a few minutes, Clay abruptly turns around and points his board toward the shore. His body goes taut and he starts to push backward. The wave is still invisible but Clay is already searching for the perfect position. And then it appears: a six-foot wall of blue. The water rises until it starts to collapse, which is when Clay pops up onto the board. He accelerates ahead of the break—his sudden speed makes the wave seem slow—and then he snaps upward, launching his board into the air and somehow whipping it around so that he lands backward on the disintegrating lip. For a dramatic moment, Clay seems off balance, but then he reverses the board and calmly rides the whitewash until it can no longer carry him. The wave is over. He's already looking for the next one.

I've just watched Clay execute the Marzo reverse, a move he pioneered. At first, these six seconds of aquatic choreography seem like every other professional surfing move: there's a whirl of saltwater spittle, a reckless leap into the air, and an improbable landing. But what makes the move so astonishing is that Clay reverses himself in midflight, swinging his surfboard in a circle and landing on top of the wave backward, facing away from the shore.

"You have no idea how hard this is," says Mitch Varnes, Clay's manager. "He is surfing the wrong way, which is *crazy*. Boards were designed to go in one direction. He makes his go in two."

This ability to invent a new move—to improvise in the sea—is a defining feature of Clay's surfing style. "I remember seeing those first grainy videos of him surfing," says Strider Wasilewski, the surf team manager at Quiksilver, Clay's surfing sponsor. "And it was clear he was doing stuff that nobody else was. He was doing moves that didn't even have names. You know how rare that is? To see a young kid doing something completely new in the water?"

This creative talent has helped make Clay one of the most celebrated surfers in the world. He already has a national surfing title and numerous Hawaiian titles; he's been featured on the cover of *Surfer* magazine and is a mainstay on YouTube. "Clay's kind of a surfing freak," says Kelly Slater, an eleven-time Association of Surfing Professionals world champion. "He's like a cat, always landing on his feet. He definitely knows things that I don't know." When I ask other pro surfers about Clay, they also mention his insatiable creativity, the way he's always improvising in the water. "Every once in a while, a surfer comes along who makes everyone else look a little boring," Strider says. "After you watch Clay, it can seem like the other guys are just performing the same moves, over and over again."

What makes Clay's innovative surfing even more impressive is that he was born with a handicap: Clay has Asperger's syndrome, a high-functioning form of autism. The syndrome is largely defined in terms of social deficits, which means that Clay is easily overwhelmed by other people and often struggles to express himself. In recent years, autism has also been associated with the absence of imagination; people with the disorder are said to suffer from

a severe literal-mindedness. In fact, the *DSM IV*—the 886-page
diagnostic guide for psychiatrists—lists "a lack of normal creativity" as one of the nine defining features of autism-spectrum disorders.

The creativity deficit of autistics has been studied most closely
by Simon Baron-Cohen, a scientist at Cambridge University. In
one study, he gave a group of autistic children a booklet of papers filled with incomplete "squiggles." Baron-Cohen then instructed the autistic children to complete the drawings by adding
new lines. "I want you to make lots of different things," he told the
kids. "Be as creative as you can."

The results of this creativity task were sobering: children with
autism scored far below normal. They completed 75 percent fewer
sketches than the control group. ("But it's not a picture of anything!" was a common response.) Baron-Cohen concluded that
children with autism suffer from a severe "imaginative deficit,"
which confined them to a more "reality-based view of the world."
They were so immersed in the literal—in the actual contours of
the squiggly lines—that they couldn't see the possibilities.

But Clay is an important reminder that there are many different forms of creativity, and that autism doesn't impair every kind.
In fact, Clay's ability to innovate in surfing is rooted in a defining
feature of his mental disorder. Hans Asperger, the Viennese pe-
diatrician who first identified the syndrome, in 1944, said that As-
pergerian children tended to have an "encompassing preoccupa-
tion with a narrow subject . . . which comes to dominate their life."
Some children with the syndrome become obsessed with nineteenth-century trains or drip coffeemakers or *The Price Is Right*.
Others will memorize camera serial numbers, even if they show
little interest in photography. Asperger argued that such obses-
siveness is often a prerequisite for important achievement, even if

it comes at a steep social cost: "It seems that for success in science and art a dash of autism is essential," Asperger wrote. "The necessary ingredient may be an ability to turn away from the everyday world . . . with all abilities cannibalized into the one specialty."

What makes Clay so unique is that his obsession is a sport and not an abstract intellectual category. He just turned twenty-two, but he can't remember a time when he wasn't obsessed with barrels, short boards, and the daily swell report. Sometimes Clay loses track of time in the water, so he ends up surfing for eight hours straight; his girlfriend has to carry fresh water to him or he gets lightheaded from dehydration. When I ask Clay what he would do if he couldn't surf, he looks confused for a second, as if he's unable to imagine such a terrifying possibility. "I don't know," he says. "I guess then I would just want to surf."

Clay was different from the start. He walked without ever having crawled: at the age of seven months and one week, he stood up and took his first steps. Although Jill, Clay's mother, is a warm and physical person—she's a professional masseuse with an easy laugh—Clay never liked being touched. He stopped nursing after only a few months. And then there was his strange relationship with water. The only way to get Clay to fall asleep as a baby was to put him in a warm bath. "We'd do four tubs a day," Jill says. "I'd put my hand under his back and let him float with the tap running. He'd pass right out."

When Clay was a year old, he started to ride on the front of his father's surfboard. Six months later, Clay was playing by himself in the shore break. At the age of two, Clay started boogie boarding; he was standing up on the board a year later. By the time he was five, Clay was riding his own short board. Before long, the ocean became his escape. "It got to the point where he'd only come out of the ocean to eat," Jill says. "He'd scarf the food down and then go right back to the beach." Late at night, Jill would often hear a commotion coming from Clay's room. She'd find him standing up in his sleep, his arms extended as if he were riding a barrel.

This obsessive interest in surfing has allowed Clay to expand the possibilities of his sport. Because he knows exactly how the ocean will behave, he can experiment with strange new moves; Clay's creativity is rooted in his deep understanding of saltwater rushing toward the shore. In other words, the narrowness of his passion has become a cognitive gift. While Clay used to resent his autistic symptoms, those social failings that got him ridiculed in school, he now sees that his burden is also a blessing. "If I didn't have Asperger's, then I wouldn't be out there [in the ocean] as much," he says. "Because I love surfing, I can completely focus on it. And so I do things in the water that maybe others can't."

Nobody knows why Clay feels so different in the saltwater. Some speculate that it's the negative ions, or the predictable rhythms of the swells, or the way liquid wraps around the body. Clay himself describes the ocean as a kind of psychological vacation, the only place where he can truly relax and be himself. "When I'm there [in the ocean] I don't need to think so much," he says. "I don't need to worry. I'm just there." The effect of the water is instant: as soon as Clay slides into the water and paddles forward on his board, all of his autistic anxieties disappear. His body goes slack and loose. It's as if his DLPFC—the fold of brain that keeps us in handcuffs—is soothed by the ceaseless pulse of the waves. The boy who is so stiff on dry land turns into a jazz musician, able to translate his immense surfing knowledge into bursts of athletic creativity. He's no longer thinking about what he doesn't understand, or what might go wrong, or what the people on shore will say. Instead, he's absorbed in the possibilities of the moment, in the strange range of movements that can be wrung from a fiberglass board and a wall of water. "People always ask me, 'How did you do that? Why did you do that?' But I never know," Clay says. "Because it's like when I'm in the water, I'm just doing it. I'm not thinking. That's why I love it."

The mysterious ability of the ocean to silence Clay's inhibitions is apparent in conversation. While spending time with Clay on Maui, I watched him flail for words during our interviews; he avoided eye contact and stared instead at his feet. Even the simplest questions led to awkward silences or stammers, as if Clay were terrified of saying the wrong thing. And yet, when I talked to Clay in the warm Hawaiian water, he surprised me with his eloquence. His sentences were filled with vivid metaphors; the same teenager who couldn't finish a sentence on dry land morphed into a poet. One day, when Clay and I were floating on our boards on a remote Maui beach, I asked him what it felt like to surf inside a barrel. "It's like being inside a throat when someone coughs and spits you out," he said. I then asked Clay what he loved about waves, why he always wanted to be in the ocean. Clay went silent and looked away at an incoming swell. I assumed he was going to ignore my question. But then Clay uttered a line that could easily be his slogan: "Waves are like toys from God. And when I'm out here, I'm just playing."

3.

The Second City theater and training center in Los Angeles—the largest school of improv in the world—is located on a seedy stretch of Hollywood Boulevard. It's surrounded by lost tourists, chintzy souvenir shops, and X-rated-movie theaters. There's the

Walk of Fame on the sidewalk, but most of the celebrity plaques are obscured by old gum. Step inside the school, however, and everything changes. The hallways echo with laughter and loud voices; hipsters run about like preschoolers on a playground.

When Second City was founded, in 1959, it was the first American theater dedicated to comic improv. At the time, the concept seemed absurd: Why would anyone want to watch actors make stuff up on stage? Performers were supposed to recite their lines, not invent their own. It didn't help that the techniques of Second City were borrowed from a set of theater games for children developed by the social worker Viola Spolin. But the childishness was the point: the premise of Second City was that little kids shouldn't have all the fun. The activities that were so liberating for third-graders might also help professional actors. And so the opening lines of Spolin's influential 1963 textbook *Improvisation for the Theater* became the credo of Second City: "Everyone can act. Everyone can improvise. Anyone who wishes to can play in the theater." It's just a matter of learning *how* to play.

In the decades since, Second City has become a factory for comic talent. Its alumni include many of the most influential figures in American comedy: John Belushi, Dan Aykroyd, John Candy, Joan Rivers, Harold Ramis, Mike Myers, Chris Farley, Steve Carell, Stephen Colbert, Adam McKay, and Tina Fey. These performers have headlined network sitcoms and pioneered classic skits on *Saturday Night Live*. They've anchored shows on Comedy Central and starred in Hollywood blockbusters. In recent years, their emphasis on improv—the kind of spontaneous humor mastered at Second City—has redefined American comedy. If the old model for funny movies was *Airplane!* or *The Naked Gun*—scripts filled with punchy one-liners and elaborate gags—the most successful comedies are now full of improvised scenes. Just look at *The Colbert Report* or *The Office* or the films of Judd Apatow: the funniest moments are rarely written in advance.[3] As a result, the jokes feel visceral and unstudied, ripe with the humor of real life. (Apatow has said that improv helps "get the imagined typer out of the way.") Bill Murray, another Second City alum, once explained why every actor should learn how to improvise: "I think that good actors always—or if you're being good, anyway—you're making it better than the script. That's your fucking job. It's like, Okay, the script says this? Well, watch *this*. Let's just roar a little bit. Let's see how high we can go."[4]

And this is why I'm visiting the Second City training center. Comic talent seems so innate—some people are just funnier than others—but the premise of these packed classes is that humor can be taught. Of course, this doesn't mean the process is going to be painless or that any of us will turn into Stephen Colbert after a few classes. "The hardest part of teaching comic improv is that people think it's so easy," says Joshua Funk, the artistic director of Second City in Hollywood. "They just see some people talking and being funny and they're like: 'I know how to be funny! I said something funny last week! I can do this!' But they can't. It

....

3. *One of my favorite examples of cinematic improv comes from the Austin Powers series. Mike Myers, playing Dr. Evil, repeatedly shushes his adult son, Scott. After a few standard shushes, Myers starts improvising, and this is when the laughs begin: "Let me tell you a story about a man named Sh!" Myers says. Scott looks confused, and then begins to open his mouth. "Shush even before you start," Myers says. He then shushes again: "That was a preemptive Sh!," before ending the scene with the following ad lib: "Just know I have a whole bag of Sh! with your name on it." The shush scene is the best in the movie, and Myers made it up on the spot.*
4. *The rise of improv in Hollywood comedies is reflected in the amount of tape that's now shot during production. While most scripted comedies shoot fewer than four hundred thousand feet of tape—a finished movie is between eight thousand and ten thousand feet long—films that rely on improv typically require more than a million feet. Apatow might use a thousand feet of film on a single scene while he lets the actors explore and improvise.*

takes years of work before you can get good at improv. It's like music that way. You can't just pick up a sax and expect to be Coltrane. You have to work at not giving a fuck."

This ability to not care what others think is one of the fundamental skills taught at Second City. A typical class begins with a series of warm-up exercises. First, the performers play a few children's games, such as duck, duck, goose or Simon says or zip-zap-zop. Then the actors start strutting around the room making a series of inappropriate bodily sounds, from errant belches to flamboyant farts. After that, the students move to "conducted rants," screaming at the top of their lungs about something that makes them angry. Finally, they arrange folding chairs in a circle and engage in what's known as "five minutes of therapy." The goal is to get confessional as quickly as possible, to share secret thoughts and repressed feelings. One student talked about a random hookup that went badly; another described a recent fight with her mother. The short stories feel raw and unfiltered, and that's the point. "The warm-ups are all about getting rid of the censor," says Andy Cobb, a veteran Second City instructor. "It's about putting people in a state of mind where they're going to say the first thing that pops into their head, even if it seems silly or stupid. Because that inner voice, that voice telling you *not* to do something—that's the voice that kills improv."

The voice Cobb is referring to comes from the dorsolateral prefrontal cortex, that chunk of tissue responsible for impulse control. The students are learning how to switch it off, how to silence their inhibitions like those jazz musicians did in the brain scanner. "When someone is struggling onstage, when they're just not being very funny, we call that 'being in your head,'" Cobb says. "The audience can smell fear. It can sense even the slightest hesitation. And that's why we spend so much time teaching people how to ignore their better judgment when they're onstage." (Keith John-stone, one of the most influential instructors of improv, puts it this way: "In life, most of us are highly skilled at suppressing action. All the improvisation teacher has to do is reverse this skill and he creates very 'gifted' improvisers.") According to Cobb, the best improv artists get so good at turning off their filters—he refers to this as "leaving your mind"—that after the show is over, they have no recollection of what just happened. "It's really weird," he says. "You're so in the moment that you disappear. You're making people laugh, but it's like it's not even you."

Of course, it's not enough to just relinquish self-control. Even as the improv actors silence their DLPFC, they still have to re-main profoundly aware of everything happening on stage. Comic improv, after all, is an ensemble performance: every joke is built on the line that came before. This is why, after the Second City students learn to stop worrying about saying the wrong thing, they begin practicing a technique called "Yes, and . . ." The ba-sic premise is simple: When performing together, improvisers can never question what came before. They need to instantly agree—that's the "yes" part—and then start setting up the next joke. I saw this process at work during a late-night performance at Second City in which six actors were doing a long-form sketch known as a Harold. For this sketch, the audience shouts out two difficult themes, and these are then woven into a single skit; in this case, the themes were obesity and religion. For a few brief seconds, the actors looked lost—they had no idea what to do. But then someone moved a chair to the center of the stage and began acting out a meal. Someone else followed with a second chair and then moaned: "Jesus, I'm hungry!" This led another actor to say: "Hold your horses, Jesus is coming as fast as he can." At this point, an actor named Jamison strolled onto the scene: "Okay, I'm here, we can all eat! And I brought some wine I just made out of water." After a few more exchanges, the story of the sketch became clear: the ensemble was acting out the Last Supper. A rambling conversation ensued, full of witty references ("Why is damn Judas always late?") and blasphemous jokes ("Jesus, these crackers are the same color as your flesh! Isn't that weird?") At one point, Jesus sees Mary Magdalene stroll by and makes a few ribald comments to Paul and Peter. The audience erupted in laughter. Not all of the lines worked, of course. Sometimes, the crowd just chuckled politely. But the sketch was often deeply funny, eliciting the kind of belly laughs that can only come from spontaneous jokes. Afterward, Andy Cobb gave them the ultimate compliment: they had been playing out of their minds.

4.

The lesson of letting go is that we constrain our own creativity. We are so worried about playing the wrong note or saying the wrong thing that we end up with nothing at all, the silence of the scared imagination. While the best performers learn how to selectively repress their inhibitions, to quiet the DLPFC on command, it's also possible to lose one's inhibitions entirely. The result is always tragic, but it's a tragedy often limned with art.

Anne Adams was a forty-six-year-old cell biologist at the University of British Columbia when she was overcome with the desire to paint. She had no artistic training or experience, just a sudden need to create. And so she bought some stretched canvases and turned a spare bedroom into her studio. Before long, Adams was spending ten hours a day making art, painting everything from local streetscapes to abstract representations of pi. After a few years, Anne began receiving high-profile commissions; her art was featured in numerous galleries and exhibitions. She continued to create for the next fifteen years, until she was felled by an incurable brain disease.

John Carter decided to become a painter shortly after his wife passed away. He was fifty-two years old and a successful investment broker with a two-stroke golf handicap. His friends were shocked by the sudden career change, as John had shown no previous interest in art. (He'd never even been to the local museum.) But John said he had no choice: he was suddenly "bombarded" by visions. And so he moved into a run-down loft, stopped eating red meat, and started wearing bright purple shirts. At first, his paintings were ugly, filled with random streaks of color. But then, after a few months of intense art-making, John's canvases started to take on an ethereal beauty. Bruce Miller, his neurologist, describes the shift:

> No one can remember exactly when John's paintings began to appeal to the eye, but it seemed to happen around the same time that he began to have trouble remembering the meaning of words. Unexpectedly, as John's language abilities eroded, his visual senses became more acute. John devoted these new visual skills to his painting, spending hours in front of a canvas perfecting every line, often using the same purples and yellows that he favored in his clothes.

John started winning numerous local art awards, and his work was displayed in a New York City gallery. Despite this artistic success, John's mental health deteriorated. His short-term memory vanished; he was prone to angry outbursts; he was no longer able to live alone. John was put on numerous psychiatric medications, but nothing helped. By the time he died, at the age of sixty-eight, he couldn't talk or drive or eat. But he still painted every day.

Anne and John both suffered from the same fatal disease:

frontotemporal dementia. Nobody knows what causes the illness, but the damage is irreversible. The disease begins when spindly neurons in the prefrontal cortex start to die; the brain area is soon riddled with holes. While frontotemporal dementia comes with a long list of terrible symptoms, from memory loss to paralysis, one of the first common effects is an insatiable need to create.

Bruce Miller is an expert on frontotemporal dementia, and his case reports are filled with patients who began losing their minds only to discover their art. There's Joan, a middle-aged housewife who stopped cooking and cleaning so that she could spend all day sketching scenes from her childhood; Mark, a wealthy business-man who made intricate wax figurines; and Miguel, a janitor who created paintings of Native American rituals in New Mexico. The details of each story are different, but the plot remains the same: These patients are suddenly overcome with the desire to paint and draw and sculpt. They lose interest in everything else. Then, after they have a few precious years of ecstatic productivity, the disease that inspired their art destroys their brains.

Why does such a devastating illness lead to a flood of creativity? The answer returns us to the prefrontal cortex, which holds back our imaginative murmurs. In frontotemporal dementia, this brain area is destroyed at a frightening pace. As a result, nothing is repressed: the raw perceptions processed in the right tempo-ral lobe of the cortex—an area devoted to multisensory integra-tion—are suddenly unleashed into the stream of consciousness. The art-making is an attempt to channel this new reality.

This might sound like a surreal condition—the mind at its most abnormal—but it's actually a condition that each of us ex-periences every night. Once we fall asleep, the prefrontal cortex shuts itself down; the censor goes eerily quiet. Meanwhile, neu-rons all across the brain start shooting out squirts of acetylcholine.

But this isn't the usual excitement of reality; this activity is semi-random and unpredictable. It's as if the mind is entertaining itself with improv, filling nighttime narratives with whatever spare details happen to be lying around.

The question, of course, is why we dream. Why does the brain squander so much energy on these nonsensical dramas? While the precise function of dreams remains unclear, there's increasing scientific evidence that they enable our creativity, allowing us to make all sorts of surprising connections. Take a 2004 paper published in *Nature* by the neuroscientists Ullrich Wagner and Jan Born. The researchers gave a group of students a tedious task that involved transforming a long list of number strings into a new set of number strings. Wagner and Born designed the task so that there was an elegant shortcut, but it could only be uncovered if the subject had an insight about the problem. When people were left to their own devices, less than 20 percent of them found the shortcut, even when given several hours to mull over the task. The act of dreaming, however, changed everything: after people were allowed to lapse into REM sleep, nearly 60 percent of them were able to discover the secret pattern. Kierkegaard was right: Sleeping is the height of genius.

Or consider a recent paper published by Sara Mednick, a neuroscientist at the University of California, San Diego. She gave subjects a variety of remote-association puzzles. Then she instructed them to take a nap. Interestingly, subjects who started to dream during their nap solved 40 percent more puzzles than they had in the morning, before their brief sleep. (Subjects who quietly rested without sleeping showed a slight decrease in performance.) According to Mednick, the reason dreams are such an important source of creativity is that, once the uptight prefrontal cortex turns itself off, we are exposed to a surfeit of surprising connections and strange ideas. Most of these new ideas will be useless, of course, just the surreal babble of the dreaming brain. But sometimes, if we're lucky, we'll find our answers in the middle of the night.

This is what happened to Keith Richards, the lead guitarist of the Rolling Stones. One night in May of 1965, Richards fell asleep early, passing out with a guitar and a tape recorder in his hotel bed. "When I wake up the next morning, I see that the tape [in the recorder] is run to the very end," Richards told Terry Gross in a 2010 interview. "And I thought: Well, I didn't do anything. You know, maybe I hit a button when I was asleep. So I put it back to the beginning and pushed play and there, in some sort of ghostly version, is the opening of a song. The whole verse is on the tape, followed by forty minutes of me snoring." The song that Richards imagined that night was "(I Can't Get No) Satisfaction," one of the most influential rock songs of all time.

The neuroscience of sleep reminds us that, every night, we temporarily turn into improv artists. Once we start to dream, we stop worrying about truth or logic or common sense. Instead of deleting our errant thoughts, we embrace the sheer freedom of our associations. And so we amuse ourselves with a set list of made-up stories and melodies, finding new connections amid the confusion.

The tragedy of frontotemporal-dementia patients is that their illness has no cure. Before long, their higher cognitive functions will start to flicker and fade. Nevertheless, this awful affliction comes with an uplifting moral, which is that all of us contain a vast reservoir of untapped creativity. The desire to make something beautiful, to express our luminous sensations, is not a rare drive confined to those with artistic training. That same desire is present in cellular biologists and stockbrokers, janitors and housewives. We don't notice this need because we constantly suppress it, because the timid circuits of the prefrontal cortex keep us from risking self-expression.

Allan Snyder, a neuroscientist at the University of Sydney, has spent the last few years documenting all this untapped creativity. He relies on a tool called transcranial magnetic stimulation, or TMS, which can temporarily silence a specific circuit of the brain with a blast of magnetic energy. Snyder is most interested in the possibility of enhancing creativity by shutting off certain brain areas, just like a jazz pianist inhibiting their DLPFC. (He refers to TMS as a "creativity-amplifying machine.") In recent years, Snyder has used the tool to selectively switch off the left frontal and temporal lobes for minutes at a time. He then asks the subject to perform a variety of activities, such as drawing an animal or solving creative puzzles. When people are hooked up to the machine, nearly 40 percent of subjects exhibit strange new talents. For instance, in the induced-drawing experiment, Snyder gives people one minute to draw an animal or face from memory. Before subjects are treated with TMS, most of their drawings are crude stick figures that don't look very much like anything. However, after people receive their "creativity treatment," their drawings are often transformed; the figures are suddenly filled with artistic flourishes. (One subject confessed that he "could hardly recognize the drawings as his own even though he had watched himself render each image.") According to Snyder, the explanation for this remarkable effect returns us to the inhibitory mechanisms of the mind, which constantly hold back our latent talents.

Picasso once summarized the paradox this way: "Every child is an artist. The problem is how to remain an artist once we grow up." From the perspective of the brain, Picasso is exactly right, as the DLPFC is the last brain area to fully develop. This helps explain why young children are so effortlessly creative: their censors don't yet exist. But then the brain matures and we become too self-conscious to improvise, too worried about saying the wrong thing, or playing the wrong note, or falling off the surfboard. It's at this point that the infamous "fourth-grade slump" in creativity sets in, as students suddenly stop wanting to make art in the classroom.

Of course, this doesn't mean that we'd be better off without our frontal lobes—we need these neural circuits to function. Nevertheless, every mental talent comes with a tradeoff. Once we learn to inhibit our impulses, we also inhibit our ability to improvise. And this is why it's so important to practice letting ourselves go.

Take this clever experiment, led by the psychologist Michael Robinson. He randomly assigned a few hundred undergraduates to two different groups. The first group was given the following instructions: "You are seven years old, and school is canceled. You have the entire day to yourself. What would you do? Where would you go? Who would you see?" The second group was given the exact same instructions, except the first sentence was deleted. As a result, these students didn't imagine themselves as seven-year-olds. After writing for ten minutes, the subjects in both groups were then given various tests of creativity, such as trying to invent alternative uses for an old car tire, or listing all the things one could do with a brick. Interestingly, the students who imagined themselves as young kids scored far higher on the creative tasks, coming up with twice as many ideas as the other group. It turns out that we can recover the creativity we've lost with time. We just have to pretend we're little kids.

Yo-Yo Ma echoes this idea. "When people ask me how they should approach performance, I always tell them that the professional musician should aspire to the state of the beginner," Ma says. "In order to become a professional, you need to go through years of training. You get criticized by all your teachers, and you worry about all the critics. You are constantly being judged. But if you get onstage and all you think about is what the critics are going to say, if all you are doing is *worrying*, then you will play terribly. You will be tight and it will be a bad concert. Instead, one needs to constantly remind oneself to play with the abandon of the child who is just learning the cello. Because why is that kid playing? He is playing for pleasure. He is playing because making this sound, expressing this melody, makes him happy. That is still the only good reason to play."

5 THE OUTSIDER

*Despite a lack of natural ability, I did have the one
element necessary to all early creativity: naïveté, that
fabulous quality that keeps you from knowing just
how unsuited you are for what you are about to do.*
—Steve Martin, *Born Standing Up*

DON LEE'S CREATIVE journey began with a broken heart. In
the winter of 2005, Don was a computer programmer for a large
insurance firm. He spent nine hours a day in a Manhattan sky-
scraper staring at a flickering computer screen and writing code.
"It was a pretty nice life," he says. "After work, I'd hang out with
the cat, sit on the couch with my girlfriend, watch some television."

But then the girlfriend left—"She even took the cat," Don
says—and he found himself with lonesome evenings and too
much free time. His empty apartment made him sad. And so Don
turned to a substance that's long been a Band-Aid for the broken-
hearted: alcohol. He started frequenting a local bar after work,
sitting at the counter and ordering a few stiff drinks. "I never went
to get drunk," Don says. "I would sip very slowly. I guess I just
wanted to be around other people."

At first, Don didn't know what to order, so he watched the bartenders work. He quickly grew enchanted with their cocktail rituals and the exacting way they crushed the ice cubes and squeezed the lemons and measured the shots. And then Don started paying attention to the taste of the drinks as he tried to memorize the subtle flavors of all the expensive liquors. He began to appreciate the smoky haze of rye, and the peaty tang of good Scotch, and the medicinal taste of gin. And then he watched as the drinks came together, as the alcohols were shaken and stirred and strained and poured. "I had nothing else to do but observe," Don says. "And I had no one else to talk to, so I talked to the bartenders."

Don was lucky: his local bar was the Pegu Club, a swank New York lounge widely celebrated for the quality of its (fifteen-dollar) cocktails. Audrey Sanders, the club's chief mixologist, was intent on bringing back classic drinks, finding inspiration in the recipes of vintage cocktail guides. Instead of listing neon green appletinis and margaritas from a mix, Sanders's menu featured obsolete libations like the gin-gin mule and the burra-peg and the Tom and Jerry. "It was only later that I realized how lucky I'd been to learn at the Pegu Club," Don says. "I learned about the importance of perfect technique and perfect ingredients. I learned about the importance of tasting and then tasting again. Those first drinks [at the Pegu Club] taught me what drinks are supposed to be."

After six months of educational imbibing, Don decided that he was ready to start making his own cocktails. Although he had virtually no bartending experience, he was asked to cover a few shifts at Death and Company, a new speakeasy in the East Village. (Brian Miller, the head bartender, had been charmed by Don's enthusiasm.) After a few months, the bartending shifts multiplied, so Don was working five nights a week while still holding down his job at the insurance firm. His routine was exhausting: he'd put in a full day of programming and then take the subway downtown and pour drinks until two in the morning. Then he'd break the bar down and catch a cab home, slinking into bed around four in the morning. After a few months at Death and Company, Don was asked to help run PDT (aka Please Don't Tell), a hip basement bar that could be entered only via a phone booth in the back of a hot-dog joint. "At first, the bar was just famous for the weird entrance and all the secrecy," Don says. "But then we started getting serious about the drinks. I wanted to develop a bar that was all about technique, so that we'd serve classic cocktails better than anyone else."[1] Although Don was still working at the insurance firm, his bartending renown was spreading. Cocktail snobs started waiting in line outside the phone booth.

Don, however, was getting restless: there was something tedious about his pursuit of alcoholic perfection. "I guess I realized that I didn't want to spend the rest of my life making the same drinks over and over," he says. "I could make an excellent martini, but it was still just a martini, you know? At a certain point, it didn't matter if I was using the best gin, the best vermouth, and the best olives. I was still just following someone else's recipe. There was nothing new. It was kinda boring."

But Don wasn't bored for long. His frustration with the strict traditions of bartending—there was only one way to make everything—soon gave way to a creative epiphany. He remembers the moment clearly: "I was prepping the bar before we opened. Cutting limes, making syrup, that sort of stuff. And I was looking up at all our bottles, all these beautiful bottles behind the bar, and I suddenly realized that I could invent a new drink. I could come

1. *The work paid off: in July of 2009, at the Cocktail Spirit Awards, PDT was voted the best cocktail bar in the world.*

up with my own cocktail. It sounds so obvious now, but for me it was a pretty big idea."

Don immediately went to work on his new cocktail project. He turned Monday and Tuesday at PDT into experimental nights during which he tested his strange new concoctions on customers. Most of the experiments were utter failures, like his attempt to carbonate a cherry. "Wouldn't it be cool if you got a Manhattan but the maraschino cherry was fizzy inside?" Don asks. "The only problem was that the fruit kept on exploding." Another experiment involved a gel cap that propelled itself around the drink—"Like a little submarine," Don says—and kept the cocktail properly mixed. Unfortunately, the propulsion system altered the flavor of the drink; the martinis tasted like baking soda.

But Don refused to get discouraged. "I was having so much fun," he says. "I was like a little kid in a candy store, except my candy was ninety proof." And so he continued to experiment, searching for new techniques and ingredients to work into his avant-garde cocktails. For Don, the research was a chance to combine his long-standing interest in chemistry—he had been an engineering major at Columbia University—with his new bartending obsession. "I was really a novice behind the bar," Don says. "Unlike the bartenders, I didn't have an encyclopedic knowledge of vodkas or single-malt whiskeys. I knew much less about liquor per se. But I knew a fair amount of chemistry, and that let me think about bartending in a slightly different way."

Look, for instance, at one of Don's first successful inventions: the bacon-infused old-fashioned. The drink relies on a process called fat-washing, in which a fatty food (like cooked bacon) is combined with an alcohol. The mixture is then chilled—the greasy globules solidify on the surface—and strained, so that no lard remains in the liquid. While fat-washing might seem like a weird concept, Don saw the experiment in terms of its elementary chemistry. "I was pretty confident it would work," he says. "Alcohol and fat have very particular atomic properties." Don then launches into a lecture on chemical polarity, a phenomenon that's caused by the separation of electrical charges within a molecule. "Polarity is why oil and water don't mix," Don says. "The fat is nonpolar and the water is polar, and so the molecules stay far apart from each other. I knew that the same principle applies to alcohol, which is also polar." However, the flavorful compounds embedded in the fat—those stray molecules that give the bacon its savory taste—are polar too, which is why they easily dissolve in the bourbon.

It took Don a few weeks of careful tinkering—he had to adjust the ratio of lard to alcohol, and then figure out how long to refrigerate the mixture—but he eventually developed a booze that made him drool. "It was a delicious drink, a perfect combination of those oaky bourbon flavors with the saltiness of breakfast meat," he says. "Unfortunately, it was a little much straight, even with ice." And so Don began testing out cocktail recipes, whisking his flavored bourbon with a variety of mixers. After adding a dash of bitters for balance, he started searching for a sweetener, as the standard squirt of simple syrup just seemed too, well, simple. His breakthrough arrived at breakfast: "I was eating pancakes and I thought about how when you pour the syrup on the pancakes and you get some maple syrup on your bacon by accident . . . That's pretty fucking tasty. So I decided to try maple syrup with the bacon-bourbon instead of straight sugar. And it worked. It worked really, really well."

Don added the bacon-infused old-fashioned to the PDT cocktail menu, serving it to his bravest customers on the experimental nights. It was an instant hit. The fat-washed bourbon soon became the bar's signature drink, garnering praise in *Gourmet, Saveur,* and *New York.* Don is now the chief mixologist at the Momofuku restaurants in New York, run by the chef David Chang. His latest cocktails make the bacon-infused old-fashioned seem conventional. There is the celery nori, in which Don steeps dried seaweed in apple brandy and then removes the seaweed and adds a dash of celery-flavored simple syrup. "I don't know how this drink works, but it does," he says. "Sometimes, you have to suspend your better judgment and just taste it." Or consider his sesame-candy cocktail, which involves mixing cognac with toasted sesame seeds and a burnt-caramel syrup. Or his pickled-ramp martini, in which tangy onion juice takes the place of olive brine. And then there is Don's clever riff on rum and Coke, which begins by fat-washing white rum with melted butter and then steeping freshly popped popcorn in it. The drink is finished with a dash of Coke. "I call it the 'movie theater,'" he says. "On the one hand, it's composed of really familiar flavors. On the other hand, it's bizarre to taste them all together in a liquid. I like inventing stuff that's just weird enough to make you think."

The success of Don Lee is a story of creativity coming from an outsider, a person on the fringes of a field. It's a parable about the benefits of knowing less—Don was a passionate amateur—and the virtues of injecting new ideas into an old field. After all, when Don invented the bacon-infused old-fashioned, he wasn't a cocktail expert. He hadn't taken any fancy bartending classes or mastered the subtleties of Kentucky whiskey or studied the history of the old-fashioned. (In fact, he was still working as a computer programmer.) "Basically, I experimented with fat-washing because I was bored and nobody told me not to," Don says. "I'm sure most bartenders would have told me it was a terrible idea, that it would never sell, that I was wasting perfectly good bourbon. But the laws of chemistry told me it should work, so why not try? I guess my only secret is that I didn't know any better."

1.

In the late 1990s, Alpheus Bingham was a vice president at Eli Lilly, one of the largest drug companies in the world. He was in charge of research strategy, helping to manage thousands of scientists working on hundreds of different technical problems. At the time, Eli Lilly's business was booming—the company was flush with "Prozac profit"—but Bingham was starting to worry about the future. The company was throwing vast sums of money at its scientific problems, desperately trying to develop the next blockbuster drug. Unfortunately, this expensive investment was producing tepid results; Bingham was beginning to wonder if there wasn't a more efficient approach to drug research. "After spending years on a problem, we'd often end up with a solution that was so imperfect it was virtually useless," he says. "And those failures weren't cheap."

For Bingham, the most troubling aspect of the drug-development model was its complete unpredictability. He had no idea which problems were solvable and which ones weren't; he couldn't anticipate how long the questions would take to answer, or where these answers would come from. "That's what really worried me—I had no idea how to manage the R and D process," Bingham says. "I didn't know who should be working on what. And that's when I started to wonder if all these supposedly impossible technical issues were really impossible. Maybe we just had the wrong people working on them? Maybe someone else could solve them? I always assumed that you hire the best resumé and give the problem to the guy with the most technical experience. But maybe that was a big mistake?"

These troubling questions led Bingham to a radical conclusion: if Eli Lilly couldn't predict which scientists would find the answer, then it needed to ask *everyone* the question. Instead of assigning its problems to particular experts inside the company, the corporation should make the problems public. "Needless to say, this strategy broke every rule of corporate R and D," Bingham says. "Like every other company, Lilly was very secretive about its research projects, for competitive reasons," he says. "You didn't want anyone else to know what you were working on." Bingham, though, was convinced that this secrecy came at a steep cost.

And so, in June of 2001, Bingham launched a website called InnoCentive. The structure of the site was simple: Eli Lilly posted its hardest scientific problems online and attached a monetary reward to each challenge. If the problem was successfully solved, then the solver got the reward. (The money was the incentive part of InnoCentive.) "Mostly we just put up these really hard organic chemistry problems," Bingham says. "I assumed there was little competitive risk, since a lot of these technical problems had also bedeviled our competitors. Frankly, I didn't expect many of these challenges to ever get solved."

A few weeks passed. The InnoCentive site was mostly silent; Bingham thought his pilot project had failed. But then, after a month of nothing, a solution was submitted. And another. And another. "The answers just started pouring in," Bingham says. "We got these great ideas from researchers we'd never heard of, pursuing angles that had never occurred to us. The creativity was simply astonishing."

After less than a year of operation, the website had become an essential R & D tool for Eli Lilly, allowing company scientists to benefit from the input of outsiders. By 2003, the site was so successful that it was spun off from its parent company and began featuring challenges from other large corporations, such as Procter and Gamble and General Electric. "These companies did the same thing Lilly had been doing," Bingham says. "They'd post the stuff they couldn't solve, put up a little prize money. Like us, they didn't expect any useful answers. But then they'd often get the solution from some researcher living halfway around the world. It was thrilling. We felt like we'd accessed this great pool of talent."

InnoCentive has continued to expand at a rapid clip. It now features problems from hundreds of corporations and nonprofits in eight different scientific categories, from agricultural science to mathematics. The challenges on the site are incredibly varied and include everything from a multinational food company looking for a "reduced-fat chocolate-flavored compound coating" to an electronics firm trying to design a lithium-ion battery for a solar-powered computer. There are calls for a spray to protect corn stalks from insect damage, and a request for a software program that can "analyze the emotional responses of consumers in a crowded retail space." More than two hundred thousand solvers have registered on the site, people who come from every conceivable scientific discipline and more than a hundred and seventy countries.

The most impressive thing about InnoCentive, however, is its effectiveness. "When it comes down to it, the only reason companies use the site is because it works," Bingham says. "It solves their hardest problems." And this success isn't merely anecdotal. In 2007, Karim Lakhani, a professor at the Harvard Business School, began analyzing hundreds of challenges posted on the site. According to Lakhani's data, nearly 40 percent of the difficult problems posted on InnoCentive were solved within six months. Sometimes the problems were solved within days of being posted online.

Think, for a moment, about how strange this is: a disparate network of strangers managed to solve challenges that Fortune 500 companies like Eli Lilly, Kraft Foods, SAP, Dow Chemical, and General Electric—companies with research budgets in the billions of dollars—had been unable to solve. By studying how these challenges got solved, Lakhani was able to better understand the surprising success rate of InnoCentive. He could see why the online amateurs were able to answer questions that had frustrated the experienced scientists.

The secret was outsider thinking: the problem solvers on Inno-Centive were most effective when working at the margins of their fields. In other words, chemists didn't solve chemistry problems, they solved molecular biology problems, just as molecular biologists solved chemistry problems. While these people were close enough to understand the challenges, they weren't so close that their knowledge held them back and caused them to run into the same stumbling blocks as the corporate scientists. "Our results showed that when the solvers rated the problem as outside their own field, they were more likely to discover the answer," Lakhani says. "Solvers were actually bridging knowledge fields—taking solutions and approaches from one area and applying them to other different areas. We have often heard that innovation occurs at the boundary of disciplines and now we have systematic evidence that this indeed is the case."

Ed Melcarek, a seven-time solver on InnoCentive, perfectly exemplifies this finding. Although Melcarek has a master's degree in particle physics, he has never solved a physics challenge on InnoCentive. Instead, he peruses the chemistry and engineering categories on the site, searching for problems that might benefit from his expertise. A few years ago, he helped Colgate-Palmolive come up with a new way of injecting fluoride powder into tubes of toothpaste. (The old method sent plumes of fluoride dust into the factory.) Melcarek's elegant solution involved imparting an electrical charge to the fluoride while grounding the plastic tube—the particles directed themselves straight inside. "It was really a very simple solution," Melcarek told *Wired*. And yet, the same fix had eluded Colgate engineers for decades.

There is something deeply counterintuitive about the success of InnoCentive. We assume that technical problems can be solved only by people with technical expertise; the researcher most likely to find the answer is the one most familiar with the terms of the question. But that assumption is wrong. The people deep inside a domain—the chemists trying to solve a chemistry problem—often suffer from a kind of intellectual handicap. As a result, the impossible problem stays impossible. It's not until the challenge is shared with motivated outsiders that the solution can be found.

Bingham likes to tell a story that demonstrates the power of InnoCentive. It involves a company that was trying to invent a polymer with a very unique and perplexing set of chemical properties. "Nobody was optimistic that InnoCentive could help the client," Bingham says. However, after a few months, solvers on the website came up with five different solutions to the problem. "The company paid for all of the solutions," Bingham says. "They paid awards to a person who studies carbohydrates in Sweden, a small agribusiness company, a retired aerospace engineer, a veterinarian, and a transdermal-drug-delivery-systems specialist. I guarantee that they would have found none of those people within their own company. They would have found none of those people if they had done a literature search in the field of interest. They would have found none of them by soliciting input from their consultants. And they probably wouldn't have hired any of these people anyway, because none of them were qualified."

2.

The world is full of natural outsiders, except we don't call them outsiders; we refer to them as young people. The virtue of youth, after all, is that the young don't know enough to be insiders, cynical with expertise. While such ignorance has all sorts of obvious drawbacks, it also comes with creative advantages, which is why so many fields, from physics to punk rock, have been defined by their most immature members. The young know less, which is why they often invent more.

The practical advantages of youth were first identified by Adolphe Quetelet, a nineteenth-century French mathematician. Quetelet's project was simple: he plotted the number of successful plays produced by playwrights over the course of their careers. That's when he discovered something unexpected: creativity doesn't increase with experience. The playwrights weren't getting better at writing plays. Instead, the curve exhibited a steep rise followed by a long, slow decline, a phenomenon of creative output now known as the inverted U curve. According to Quetelet, his curve demonstrated that creativity tends to peak after a few years of work—when we know enough, but not too much—before it starts to fall, in middle age.

Dean Simonton, a psychologist at UC-Davis, has spent the last several decades expanding on Quetelet's approach, sifting through vast amounts of historical data in search of the subtle patterns that influence creative production over time. For instance, Simonton has shown that physicists tend to make their most important discoveries early in their careers, typically before the age of thirty. The only field that peaks before physics is poetry.

Why are young physicists and poets more creative? One possibility is that time steals ingenuity, that the imagination starts to wither in middle age. But that's not the case—we are not bio-logically destined to get less creative. Simonton argues that youth benefit from their outsider status—they're innocent and ignorant, which makes them more willing to embrace radical new ideas. Be-cause they haven't become encultured, or weighted down with too much conventional wisdom, they're more likely to rebel against the status quo.[2] After a few years in the academy, Simonton says, the "creators start to repeat themselves, so that it becomes more of the same-old, same-old." They have become insiders.

But there is nothing inevitable about this process—creativity doesn't have to slowly slip away. As Simonton notes, we can con-tinue to innovate for our entire careers as long as we work to main-tain the perspective of the outsider. Just look at the mathemati-cian Paul Erdos, who was one of the most productive scientists of all time. Erdos was famous for hopscotching around his discipline, working with new people on new problems. He embraced a multi-plicity of subjects, publishing influential papers on number theory, topology, combinatorics, and probability. At the first hint of bore-dom—and Erdos got bored very quickly—he would begin again, starting over with a new challenge and a blank sheet of paper. As a result, his creative output never declined; there was no U curve for his career, just a sharp rise followed by a flat line.[3] "If you can keep finding new challenges, then you can think like a young per-son even when you're old and gray," Simonton says. "That idea gives me hope."

....

2. This also helps explain the disconnect between education and creativity. According to Simonton's data, the ideal amount of college for a creative career is two years of under-graduate work. After that, school seems to actually inhibit the imagination. Mihaly Csik-szentmihalyi, a psychologist at Claremont, is blunter. He notes that, in most instances, "school threatens to extinguish the interest and curiosity that the child had discovered outside its walls."
3. The Benzedrine didn't hurt either.

The moral is that outsider creativity isn't a phase of life — it's a *state of mind*. Of course, it's not easy cultivating this useful mental state, at least once we get older. Sometimes we have to work a second job, mixing cocktails when we're not programming insurance software. Sometimes we have to spend our free time working on confusing problems, or immersing ourselves in strange new fields, or wasting lots of bourbon on a crazy bacon experiment. We need to be willing to risk embarrassment, ask silly questions, surround ourselves with people who don't know what we're talking about. We need to leave behind the safety of our expertise.

But sometimes that's not enough: we need to leave behind *everything*. One of the most surprising (and pleasurable) ways of cultivating an outsider perspective is through travel, getting away from the places we spend most of our time. The reason travel is so useful for creativity involves a quirk of cognition in which problems that feel close get contemplated in a more literal manner. This means that when we are physically near the source of the problem, our thoughts are automatically constricted, bound by a more limited set of associations. While this habit can be helpful — it allows us to focus on the facts at hand — it also inhibits the imagination.

Consider a field of corn. When you're standing in the middle of a farm surrounded by the tall cellulose stalks and fraying husks, the air smelling faintly of fertilizer and popcorn, your mind is automatically drawn to thoughts related to the primary definition of *corn*, which is that it's a plant, a cereal, a staple of midwestern farming. But imagine that same field of corn from a different perspective. Instead of standing on a farm, you're now in a crowded city street dense with taxis and pedestrians. The plant will no longer be just a plant; instead, your vast neural network will pump out all sorts of associations. You'll think about high-fructose corn syrup, obesity, and the Farm Bill; you'll contemplate ethanol and the Iowa caucuses, those corn mazes for kids at state fairs, and the deliciousness of succotash made with bacon and lima beans. The noun is now a web of tangents, a vast loom of connections.

And this is why travel is so helpful: When you escape from the place you spend most of your time, the mind is suddenly made aware of all those errant ideas previously suppressed. You start thinking about obscure possibilities—corn can fuel cars!—that never would have occurred to you if you'd stayed back on the farm. Furthermore, this expansive kind of cognition comes with practical advantages, since you can suddenly draw on a whole new set of possible solutions.[4]

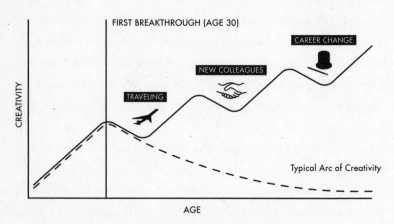

While the inverted U curve of creativity holds back most people, there are ways to remain creative over time.

....

4. *But it's not enough to simply get on a plane; if you want to experience the creative benefits of travel, then you have to rethink its raison d'être. Most people, after all, escape to Paris so they don't have to think about those troubles they left behind. But here's the ironic twist: your mind is most likely to solve your stubbornest problems while you're sitting in a swank Left Bank café. So instead of contemplating that buttery croissant, mull over those domestic riddles you just can't solve. You might have the breakthrough while on break.*

Look, for instance, at a recent experiment led by the psychologist Lile Jia at Indiana University. He randomly divided a few dozen undergraduates into two groups, each of which were asked to list as many different modes of transportation as possible. (This is known as a creative generation task.) One group of students was told that the task was conceived by Indiana University students studying abroad in Greece, while the other group was told that it was conceived by Indiana students studying in Indiana. At first, it's hard to believe that such a slight and seemingly irrelevant distinction would alter the performance of the subjects. Why would it matter where the task originated?

Nevertheless, Jia found a striking difference between the two groups: when students were told that the task was imported from Greece, they came up with significantly more transportation possibilities. They didn't limit their list to cars, buses, trains, and planes; they cited horses, triremes, spaceships, bicycles, and Segway scooters. Because the source of the problem was far away, the subjects felt less constrained by their local transport options; they didn't think about getting around just in Indiana, they thought about getting around all over the world.

In a second study, Jia found that Indiana University students were much better at solving a series of insight puzzles when told that the puzzles came from California and not from Indiana. Here's a sample problem:

> A prisoner was attempting to escape from a tower. He found a rope in his cell that was half as long as required to permit him to reach the ground safely. He divided the rope in half, tied the two parts together, and escaped. How could he have done this?

The sense of distance from where the puzzle originated allowed these subjects to imagine a far wider range of alternatives, which made them more likely to solve the challenging brainteasers. (The answer to the sample problem is that the prisoner unraveled the rope lengthwise and tied the remaining strands together.) Instead of getting stuck and giving up, they were able to think about unusual associations, which eventually led to the right answer.

The larger lesson is that our thoughts are shackled by the familiar. The brain is a neural tangle of near-infinite possibility, which means that it spends a lot of time and energy choosing what *not* to notice. As a result, creativity is traded for efficiency; people think in literal prose, not symbolist poetry. It's not until we feel distant from the problems—far from our usual haunts—that the chains of cognition are loosened, and the insight becomes obvious.

What's more, the longer you're away from home, the stronger the effect. In a 2009 study, researchers at INSEAD and the Kellogg School of Management reported that students who lived abroad for an extended period were significantly more likely to solve a difficult creativity problem than students who had never lived outside of their birth country. The experiment went like this: The students were given a cardboard box containing a few thumbtacks, a piece of corkboard, a book of matches, and a waxy candle. They were told to attach the candle to the piece of corkboard so that it could burn properly without dripping wax onto the floor. This is known as the Duncker candle problem, and it tends to make people very frustrated. In fact, nearly 90 percent of people pursue the same two failing strategies. They begin by tacking the candle directly to the board, which causes the candle wax to shatter. Then they attempt to melt the candle with the matches so that it sticks to the board. But the wax doesn't hold; the candle falls to the floor. At this point, most people surrender. They assume that the puzzle is impossible, that it's a stupid experiment and a waste of time. In fact, only a slim minority of subjects manage to come up with the solution, which involves attaching the candle to the cardboard box with wax and then tacking the cardboard box to the corkboard. Unless people have an insight about the box—that it can do more than hold thumbtacks—they'll waste candle after candle, repeating their failures while waiting for a breakthrough. Psychologists refer to this as the bias of functional fixedness, since people are typically terrible at coming up with new functions for old things.

What does this have to do with living abroad? According to the researchers, the experience of another culture endows the traveler with a valuable open-mindedness, making it easier for him or her to realize that a single thing can have multiple meanings.[5] Consider the act of leaving food on one's plate. In China, this is often seen as a compliment, a signal that the host has provided more than enough food. But in America the same act is an insult, an indication that the food wasn't good enough to finish.

Such cultural contrasts mean that seasoned travelers are alive to ambiguity, more willing to realize that there are different (and equally valid) ways of interpreting the world. Because they've felt like outsiders before, immersed in foreign places, they've learned to examine alternative possibilities. This, in turn, allows them to expand the circumference of their "cognitive inputs," as they re-

---

5. *The same principle is also true of people with multiple social identities. According to a study led by Jeffrey Sanchez-Burks, a psychologist at the University of Michigan, people who describe themselves as both Asian and American, or see themselves as female engineers (and not just engineers), display higher levels of creativity. The reason is that social identities are often associated with distinct problem-solving approaches. As a result, when these types of people are faced with a challenging puzzle, their minds remain more flexible, better able to experiment with multiple creative strategies. Pluralism is always practical.*

fuse to settle for their first answers and initial guesses. Maybe the box has a different function. Maybe there's a better way to attach a candle to a board.

Of course, this mental flexibility doesn't come from mere distance. It's not enough just to change time zones or schlep across the world only to eat Le Big Mac instead of a Quarter Pounder with Cheese. Instead, this increased creativity appears to be a side effect of experiencing *difference:* we need to change cultures, to feel the disorienting diversity of human traditions. The same details that make foreign travel so confusing—Do I tip the waiter? Where is this train taking me?—turn out to have a lasting impact, making us more creative because we're less insular. We're reminded of all that we don't know, which is nearly everything; we're surprised by the constant stream of surprises. Even in this globalized age, the world slouching toward similarity, we can still marvel at all the earthly things that weren't included in the *Let's Go* guidebook and that certainly don't exist back in Indiana. When we get home, home is still the same. But something in our minds has been changed, and that changes everything.

Ruth Handler learned this lesson firsthand. In the early 1950s, she spent countless hours watching her young daughter, Barbara, play with paper dolls. Although the cutouts looked like little children, Handler noticed that Barbara often gave her dolls adult roles. Sometimes she would play waitress with the paper figures or pretend that one of the paper children was actually a mother. And that's when Handler had her "crazy idea," which she would later describe in her memoir *Dream Doll*:

> Barbara was using these dolls to project her dream of her own future as an adult woman. So one day it hit me: Wouldn't it be great if we could take that play pattern and three-dimensionalize it so that little girls could do their dreaming and role-play-

ing with real dolls instead of the flimsy paper ones? It dawned on me that this was a basic, much needed play pattern that had never before been offered by the doll industry to little girls.

At the time, Handler's husband was an executive at the Mattel toy company. When Ruth suggested that the company create a doll that looked like an adult, he immediately rejected the idea as unfeasible and silly. Little girls didn't want to play with grownups! Besides, what did Ruth know about the toy business? She was just a mother. And so the proposal was shelved; Mattel continued to produce its line of paper dolls, all of which looked like infants.

It was a vacation that made all the difference. In the summer of 1956, the Handler family traveled to Europe for the first time. Ruth was wandering around a small town in Switzerland when she noticed a strange-looking doll in the window of a cigarette shop. The doll was eleven inches tall and had platinum-blond hair, long legs, and an ample bosom. Her name was Bild Lilli. Although Handler didn't know it at the time—she didn't speak German—the doll was actually a sex symbol, sold mainly to middle-aged men. (That's why the doll was only stocked in bars and tobacco stores.) But Handler didn't get the joke—she took one look at the blond Bild Lilli and saw a perfect toy for young girls.

When the family returned home, Handler continued to lobby her husband to build an Americanized version of the Bild Lilli. The European prototype was ultimately persuasive. After watching young girls play with the adult toy, the executives at Mattel realized that Ruth's proposal had tremendous potential. And so, in March of 1959, Mattel released the Barbie doll. Although the toy was initially a flop—Sears refused to carry a children's product with "feminine curves"—sales steadily increased. Before long, the plastic toy became a cultural icon, beloved by girls, burned by feminists, and immortalized by Warhol. Mattel has since sold more than a billion Barbies—a salacious German figurine is now one of the most popular toys in the world.

Handler saw the potential of the Bild Lilli doll only because she was an outsider. If she'd spoken German or been a local—if she'd grasped the bawdy backstory—then she never would have considered the doll for her daughter. She would have disregarded it as a tasteless product, yet more evidence that an adult doll wasn't feasible. And if she'd been a toy executive, the idea probably wouldn't have occurred to her in the first place, since everyone knew that little girls want to play with babies. But Ruth Handler didn't know any of this; her *misunderstanding* led to the insight. She was just a mother lost in a foreign country, and that's why she invented the Barbie.

3.

The outsider problem affects everyone. Although we live in a world that worships insiders, it turns out that gaining such expertise takes a toll on creativity. To struggle at anything is to become too familiar with it, memorizing details and internalizing flaws. It doesn't matter whether you're designing a city park or a shoot-'em-up video game, whether you're choreographing a ballet or a business conference: you must constantly try to forget what you already know.

This is one of the central challenges of writing. A writer has to read his sentences again and again. (Such are the inefficiencies of editing.) The problem with this process is that he very quickly loses the ability to see his prose as a reader and not as the writer. He knows exactly what he's trying to say, but that's because he's the one saying it. In order to construct a clear sentence or a coherent narrative, he needs to edit as if he knows nothing, as if he's never seen these words before.

This is an outsider problem—the writer must become an outsider to his own work. When he escapes from the privileged position of the author, he can suddenly see all those imprecise clauses and unnecessary flourishes; he can feel the weak parts of the story and the slow spots in the prose. That's why the novelist Zadie Smith, in an essay on the craft of writing, stresses the importance of putting aside one's prose and allowing the passage of time to work its amnesiac magic.

> When you finish your novel, if money is not a desperate priority, if you do not need to sell it at once or be published that very second—put it in a drawer. For as long as you can manage. A year or more is ideal—but even three months will do . . . You need a certain head on your shoulders to edit a novel, and it's not the head of a writer in the thick of it, nor the head of a professional editor who's read it in twelve different versions. It's the head of a smart stranger, who picks it off a bookshelf and begins to read. You need to get the head of that smart stranger somehow. You need to forget you ever wrote that book.

Why is it so important to get some distance from one's prose? Stanislas Dehaene, a neuroscientist at the College de France in Paris, has helped illuminate the neural anatomy of reading and editing. It turns out that the brain contains two distinct pathways for making sense of words, each of which is activated in a different context. One pathway is known as the ventral route, and it's direct and efficient, accounting for the vast majority of our literacy. The process is straightforward: You see a group of letters, convert those letters into a word, and then directly grasp the word's semantic meaning. According to Dehaene, this ventral pathway is turned on by "routinized, familiar passages" of prose and relies on a bit of cortex known as the visual word form area (VWFA). When reading a simple sentence or a paragraph full of clichés, you're almost certainly relying on this ventral neural highway. As a result, the act of reading seems effortless and easy. You don't have to think about the words on the page.

But the ventral route is not the only way to read. The second reading pathway—known as the dorsal stream—is turned on whenever you're forced to pay conscious attention to a sentence, perhaps because of an obscure word, an awkward subclause, or bad handwriting. (In his experiments, Dehaene activates this pathway by rotating the letters or filling the prose with errant punctuation.) Although scientists had previously assumed that the dorsal route ceased to be active once an individual learns how to read, Dehaene's research demonstrates that even literate adults are still occasionally forced to make sense of texts. Once that happens, they become more conscious of the words on the page.

This suggests that the act of reading observes a gradient of awareness. Familiar sentences printed in Helvetica are read quickly and effortlessly, while unusual sentences with complex clauses and smudged ink tend to require more work, which leads to more activation in the dorsal pathway. The same principle might also apply to one's own writing: when you read something that you've just composed, everything is ventral. You're so familiar with the words that you don't really notice them—it's literacy at its most automatic. The end result is that all writerly mistakes become invisible, since the brain isn't paying close attention; the bad prose has been internalized. In contrast, if you follow Zadie Smith's advice and give your sentences time to be forgotten—if you start to read as an outsider, and not as the author—then you'll rely more on the dorsal stream. This allows you to think more critically about the words on the page. You suddenly notice what you previously ignored, all those pointless metaphors and pretentious adjectives, those redundant sentences and boring paragraphs. At long last, you know how it needs to be.

Knowledge can be a subtle curse. When we learn about the world, we also learn all the reasons why the world *cannot* be changed. We get used to our failures and imperfections. We become numb to the possibilities of something new. In fact, the only way to remain creative over time — to not be undone by our expertise — is to experiment with ignorance, to stare at things we don't fully understand. This is the lesson of Samuel Taylor Coleridge, the nineteenth-century Romantic poet. One of his favorite pastimes was attending public chemistry lectures in London, watching eminent scientists set elements on fire. When Coleridge was asked why he spent so much time watching these pyrotechnic demonstrations, he had a ready reply. "I attend the lectures," Coleridge said, "so that I can renew my stock of metaphors." He knew that we see the most when we are on the outside looking in.

6 THE POWER OF Q

*Not everyone can become a great artist, but a great
artist can come from anywhere.*
—Anton Ego, in Pixar's *Ratatouille*

THE SOURCE OF every new idea is the same. There is a network
of neurons in the brain, and then the network shifts. All of a sudden, electricity flows in an unfamiliar pattern, a shiver of current
across a circuit board of cells. But sometimes a single network isn't
enough. Sometimes a creative problem is so difficult that it requires people to connect their imaginations together; the answer
arrives only if we collaborate. That's because a group is not just
a collection of individual talents. Instead, it is a chance for those
talents to exceed themselves, to produce something greater than
anyone thought possible. When the right mixture of people come
together and when they collaborate in the right way, what happens can often feel like magic. But it's not magic. There is a reason
why some groups are more than the sum of their parts.

Furthermore, there's evidence that group creativity is becoming more necessary. Because we live in a world of very hard
problems—all the low-hanging fruit is gone—many of the most
important challenges exceed the capabilities of the individual

imagination. As a result, we can find solutions only by working with other people.

Ben Jones, a professor of management at the Kellogg Business School, has demonstrated this by analyzing trends in "scientific production." The most profound trend he's observed is a sharp shift toward scientific teamwork. By analyzing 19.9 million peer-reviewed papers and 2.1 million patents from the last fifty years, Jones was able to show that more than 99 percent of scientific subfields have experienced increased levels of teamwork, with the size of the average team increasing by about 20 percent per decade. While the most cited studies in a field used to be the product of lone geniuses—think Einstein or Darwin—Jones has demonstrated that the best research now emerges from groups. It doesn't matter if the researchers are studying particle physics or human genetics: science papers produced by multiple authors are cited more than twice as often as those authored by individuals. This trend was even more apparent when it came to "home-run papers"—those publications with at least a thousand citations—which were more than *six times* as likely to come from a team of scientists.

The reason is simple: the biggest problems we need to solve now require the expertise of people from different backgrounds who bridge the gaps between disciplines. Unless we learn to share our ideas with others, we will be stuck with a world of seemingly impossible problems. We can either all work together or fail alone.

But how should we work together? What's the ideal strategy for group creativity? Brian Uzzi, a sociologist at Northwestern, has spent his career trying to answer these crucial questions, and he's done it by studying Broadway musicals. Although Uzzi grew up in New York City and attended plenty of productions as a kid, he doesn't exactly watch *A Chorus Line* in his spare time. "I like musicals just fine, but that's not why I study them," he says. Instead, Uzzi spent five years analyzing thousands of old musicals because he sees the art form as a model of group creativity. "Nobody creates a Broadway musical by themselves," Uzzi says. "The production requires too many different kinds of talent." He then rattles off a list of the diverse artists that need to work together: the composer has to write songs with a lyricist and librettist, and the choreographer has to work alongside the director, who is probably getting notes from the producers.

Uzzi wanted to understand how the relationships of these team members affected the end result. Was it better to have a group composed of close friends who had worked together before, or did total strangers make better theater? What is the ideal form of creative collaboration? To answer these questions, Uzzi undertook an epic study of nearly every musical produced on Broadway between 1877 and 1990, analyzing the teams behind 2,258 different productions. (To get a full list of collaborators, he often had to track down dusty old *Playbills* in theater basements.) He charted the topsy-turvy relationships of thousands of different artists, from Cole Porter to Andrew Lloyd Webber.

The first thing Uzzi discovered was that the people who worked on Broadway were part of an extremely interconnected social network: it didn't take many links to get from the librettist of *Guys and Dolls* to the choreographer of *Cats*. Uzzi then came up with a way to measure the density of these connections for each musical, a designation he called Q. In essence, the amount of Q reflects the "social intimacy" of people working on the play, with higher levels of Q signaling a greater degree of closeness. For instance, if a musical was being developed by a team of artists who had worked together several times before—this is common practice on Broadway, since producers see "incumbent teams" as less risky — that musical would have an extremely high Q. In contrast, a musical created by a team of strangers would have a low Q.

This metric allowed Uzzi to explore the correlation between levels of Q and the success of the musical. "Frankly, I was surprised by how big the effect was," Uzzi says. "I expected Q to matter, but I had no idea it would matter this much." According to the data, the relationships between collaborators was one of the most important variables on Broadway. The numbers tell the story: When the Q was low, or less than 1.7, the musicals were much more likely to fail. Because the artists didn't know one another, they struggled to work together and exchange ideas. "This wasn't so surprising," Uzzi says. "After all, you can't just put a group of people who have never met before in a room and expect them to make something great. It takes time to develop a successful collaboration." However, when the Q was too high (above 3.2) the work also suffered. The artists were so close that they all thought in similar ways, which crushed theatrical innovation. According to Uzzi, this is what happened on Broadway during the 1920s. Although the decade produced many talented artists — Cole Porter, Richard Rodgers, Lorenz Hart, and Oscar Hammerstein II — it was also full of theatrical failures. (Uzzi's data revealed that 87 percent of musicals produced during the decade were utter flops, which is far above the historical norm.) The problem, he says, is that all of these high-profile artists fell into the habit of collaborating with only their friends. "Broadway [during the 1920s] had some of the biggest names ever," says Uzzi. "But the shows were too full of repeat relationships, and that stifled creativity. All the great talent ended up producing a bunch of mediocre musicals."

What kind of team, then, led to the most successful musicals? Uzzi's data clearly demonstrates that the best Broadway shows were produced with *intermediate* levels of social intimacy. A musical produced at the ideal level of Q (2.6) was two and a half times more likely to be a commercial success than a musical produced with a low Q (<1.4) or a high Q (>3.2). It was also three times more likely to be lauded by the critics. This led Uzzi to argue that creative collaborations have a sweet spot: "The best Broadway teams, by far, were those with a mix of relationships," Uzzi says. "These teams had some old friends, but they also had newbies. This mixture meant that the artists could interact efficiently—they had a familiar structure to fall back on—but they also managed to incorporate some new ideas. They were comfortable with each other, but they weren't *too* comfortable."

Low Q   Ideal Q   High Q

The most creative teams aim for the sweet spot of Q.

Uzzi's favorite example of intermediate Q is *West Side Story*, one of the most successful Broadway musicals of all time. In 1957, the play was seen as a radical departure from Broadway conventions, for both its willingness to tackle social problems and its extended dance scenes. At first, *West Side Story* might look like a play with a high Q, since several of its collaborators were already Broadway legends who had worked together before. The concept for the play emerged from a conversation among Jerome Robbins, Leonard Bernstein, and Arthur Laurents. But that conversation among old friends was only the beginning. As Uzzi points out, *West Side Story* also benefited from a crucial injection of unknown talent. A twenty-five-year-old lyricist named Stephen Sondheim was hired to write the words (even though he'd never worked on Broadway before), while Peter Gennaro, an assistant to Robbins, provided many important ideas for the choreography. "People have a tendency to want to only work with their friends," says Uzzi. "It feels so much more comfortable. But that's exactly the wrong thing to do. If you really want to make something great, then you're going to need to seek out some new people too."

1.

The screenwriter William Goldman in his memoir *Adventures in the Screen Trade* famously declared that "the single most important fact of the entire movie industry [is that] NOBODY KNOWS ANYTHING." To demonstrate his point, Goldman cited a long list of Hollywood flops and surprise successes. For instance, one of the highest-grossing movies in history, *Raiders of the Lost Ark,* was offered to every studio in Hollywood, and every one of them turned it down except Paramount: "Why did Paramount say yes?" Goldman asks. "Because nobody knows anything. And why did all the other studios say no? Because nobody knows anything. And why did Universal, the mightiest studio of all, pass on *Star Wars* ...? Because nobody, *nobody* — not now, not ever — knows the least goddam thing about what is or isn't going to work at the box office." Hollywood, in other words, is like a slot machine: every movie is a blind gamble.

Pixar Animation Studios is the one exception to Goldman's rule. Since 1995, when the first *Toy Story* was released, Pixar has created eleven feature films. Every one of those films has been a commercial success, with an average international gross of more than $550 million *per film*. These blockbusters have also been critical darlings; the studio has collected twenty-four Academy Awards, six Golden Globes, and three Grammys. Since 2001, when the Oscars inaugurated the category of Best Animated Feature, every Pixar film has been nominated; five of those films have taken home the statue.

The only way to understand Pixar's success is to understand its unique creative process, which has slowly evolved over the course of its thirty-year history. Before Pixar was a movie studio, it was a computer manufacturer. The roots of the company date to 1980, when the director George Lucas started a computer division within Lucasfilm, his movie production firm. At the time, Lucasfilm was flush with profit from *Star Wars* and *The Empire Strikes Back,* and George Lucas was interested in exploring the possibility of using these new machines to create cinematic special effects. (All of the effects for *Star Wars* had been done manually; the lightsabers, for instance, were painted onto each frame of film.) And so, in 1980, Lucas hired Ed Catmull and Alvy Ray Smith, two computer scientists who specialized in the creation of digital imagery.

Although Lucas was funding this avant-garde research, he showed little interest in using special effects in his films. In fact, the first cinematic application of this new technology came in *Star Trek II: The Wrath of Khan,* when the camera swooped onto the surface of a distant planet. "That was our big break," Smith remembers. "It took a long time to make just a few seconds of film, but I was jazzed right to the teeth about what we'd done. We showed that even these slow machines could make something that looked pretty remarkable."

But Catmull and Smith weren't content to work on short digital illusions; they wanted to make their own feature film, an animated movie that would be created entirely on the computer. Unfortunately, Lucas had no intention of letting his scientists become filmmakers. As a result, Catmull and Smith had to shroud their animation project in secrecy. Their first hire was a Disney animator named John Lasseter, who was given the vague official title of user-interface designer. Catmull and Smith had been working on a short cartoon called *The Adventures of André and Wally B*—it featured a character being woken by a pesky bumblebee—and Lasseter immediately made several major changes. He replaced the rigid geometry of circles and squares with more varied shapes and injected some comedy into the interactions of André and the insect. "It was very clear from the beginning that John was a master storyteller," says Catmull, who is the current president of Disney Animation Studios and Pixar. "He had a skill set that we desperately needed. And so we basically listened to everything he had to say."

While *André* was a technical triumph—it's widely celebrated for spurring interest in computer animation among the major Hollywood studios—George Lucas was getting tired of funding a bunch of computer geeks and their expensive mainframes. Enter Steve Jobs. At the time, Jobs was still smarting from being forced out of Apple, and he saw the computer division at Lucasfilm as a potential investment. But Jobs wasn't that interested in animation. He was drawn to the Pixar Image Computer, a $135,000 machine capable of generating complex graphic visualizations. (Catmull and Smith justified their cartoons as marketing tools that showed off the power of the hardware.) In 1986, Jobs bought the computer division for $10 million from Lucasfilm. The new company was named after its only product: Pixar.

Unfortunately, the expensive computers were a commercial flop. ("We were just a little too far ahead of the curve," says Smith. "People weren't ready to spend that much money on a computer that could only produce pictures.") Jobs was forced to extend a personal line of credit to Pixar, which was losing millions of dollars every year. Meanwhile, Catmull and Smith were scrambling to bring in revenue, if only to keep their creative team together. The two scientists soon came up with a plan: they would start producing commercials. While the technology wasn't yet ready for a feature film—the computers were still too slow—the Pixar machines could efficiently render fifteen-second television spots. Before long, Lasseter was animating ads for Listerine, Lifesavers, Volkswagen, and Trident gum. It wasn't particularly fulfilling work, but it paid the bills.

Despite these financial struggles, a unique creative culture was developing within Pixar. This culture was defined by the free flow of ideas, by the constant interaction between computer scientists and cartoon animators. At first, these interactions were a byproduct of the technology, which remained so fraught with problems that each short film became an endless negotiation. Was this irregular shape possible to animate? What could be done about motion blur? How could a facial expression become more expressive? Because Pixar was inventing its own art form, every aesthetic decision had technical consequences, and these would then require more artistic tweaks. "In those early days, we had no idea what we were doing," says Bobby Podesta, a supervising animator. "We were groping in the dark. That meant we needed to constantly consult the computer guys—'Can you do this? What about this?'—and then push them when they said it couldn't be done. It became this never-ending conversation where we were all trying to figure out what was even possible."

To help sell the hardware, Pixar continued making short films. The most impressive was *Tin Toy,* a story about a wind-up toy running away from a baby. Lasseter was inspired to make the short after watching a home video of his nephew: "The video was half an hour of him just sitting there, playing with his toys," Lasseter says. "Everything he picked up went into his mouth, and he slobbered all over it. I thought, Ahh, imagine what it must be like to be a toy in the hands of a baby. That baby must seem like a monster. And that idea is where *Tin Toy* came from." *Tin Toy* was such a critical success—it became the first computer-animated film to win an Oscar—that Disney Studios decided to collaborate with Pixar on a feature film, one that would also revolve around the emotional relationship between a toy and its owner. The working title of the movie was *Toy Story,* if only because nobody could think of anything better. "This was our big break," says Catmull. "But it was also pretty intimidating. We were used to making short commercials, not an eighty-two-minute film."

At the time, Disney pressured Pixar to create a separate production company for *Toy Story.* This was standard Hollywood procedure: "Everybody told us that when you made a movie, you formed a company within a company and separated out the cultures," Catmull says. "We'd never made a movie before, so what did we know? We came up with a name"—the production company was going to be called Hi-Tech Toons—"and even printed up stationery. But then I went to John [Lasseter] and showed him the logo and stuff, and he said 'That's a really bad idea.' So we canceled the plans. We told all the Hollywood people we were going to do it our way."

The reason Pixar decided against an independent production company was that it didn't want to place any constraints on the interactions of its employees. Pixar realized that its creativity emerged from its culture of collaboration, its ability to get talented people from diverse backgrounds to work together. (Lasseter describes the equation this way: "Technology inspires art, and art challenges the technology.") While the studio was determined to hire gifted animators and ingenious computer scientists, it was just as determined to get these new hires to interact with one another and with older, more experienced employees. The meritocracy needed to mingle. Of course, the only way to cultivate this kind of collaboration — the right level of Q — was to have everyone in the same building, and not scattered among various spinoffs and independent entities. "The modern Hollywood approach was to put together a team for one project and then disband the team when production was finished," Catmull says. "But we thought that was dumb. When it comes down to it, the only way to make a good movie is to have a good team. The current view in Hollywood, in contrast, is that movies are all about ideas, and that a good idea is rarer and more valuable than good people. That's why there are so many copycat movies: everyone is chasing the same concept. But that's a fundamentally misguided approach. A mediocre team will screw up a good idea. But if you give a mediocre idea to a great team and let them work together, they'll find a way to succeed."

2.

Pixar Animation Studios is set in an old Del Monte canning factory just north of Oakland. The studio originally planned to build something else, an architectural design that called for three buildings, with separate offices for the computer scientists, animators, and management. While the layout was cost-effective — the smaller, specialized buildings were cheaper to build — Steve Jobs scrapped the plan. ("We used to joke that the building was Steve's movie," Catmull says. "He really oversaw everything.") Before long, Jobs had completely reimagined the studio. Instead of three buildings, there was going to be a single vast space with an airy atrium at its center. "The philosophy behind this design is that it's good to put the most important function at the heart of the building," Catmull says. "Well, what's our most important function? It's the interaction of our employees. That's why Steve put a big empty space there. He wanted to create an open area for people to always be talking to each other."

But Jobs realized that it wasn't enough simply to create an airy atrium; he needed to force people to go there. Jobs began with the mailboxes, which he shifted to the lobby. Then he moved the meeting rooms to the center of the building, followed by the cafeteria and coffee bar and gift shop. But that still wasn't enough, which is why Jobs eventually decided to locate the only set of bathrooms in the atrium. "At first, I thought this was the most ridiculous idea," says Darla Anderson, an executive producer on several Pixar films. "I have to go to the bathroom every thirty minutes. I didn't want to have to walk all the way to the atrium every time I needed to go. That's just a waste of time. But Steve said, 'Everybody has to run into each other.' He really believed that the best meetings happened by accident, in the hallway or parking lot. And you know what? He was right. I get more done having a bowl of cereal and striking up a conversation or walking to the bathroom and running into unexpected people than I do sitting at my desk." Brad Bird, the director of *The Incredibles* and *Ratatouille*, agrees: "The atrium initially might seem like a waste of space . . . But Steve realized that when people run into each other, when they make eye contact, things happen. So he made it impossible for you not to run into the rest of the company."

And it's not just the atrium; the atmosphere of interaction is evident all across the campus. When I visited the studio, during the final, frantic days of production on *Toy Story 3*, it seemed as if every common space echoed with conversation. There was, as

Jobs predicted, plenty of chatter inside the bathroom,[1] but there were also crowds talking in the coffee bar about the Randy Newman soundtrack, and large groups sharing jokes over plates of Thai curry at the Luxo Café. I saw people collaborating in the art gallery and listened to animators talk shop while sitting in their Barcaloungers. (In the evenings, the social activity transitions to the bars — there are eleven drinking holes on the Pixar campus.) And then there's Pixar University, a collection of 110 different classes, from creative writing to comic improv, that are offered to all employees. The classes are filled with a diverse group of students, so John Lasseter might learn how to juggle in the atrium alongside a security guard. The Latin crest of Pixar University says it all: *Alienus Non Diutius*, which means "alone no longer."

The sociologist Ray Oldenburg referred to such gathering spots as "third places," which he defined as any interactive environment that is neither the home (the first place) nor the office (the second place). These shared areas have played an outsize role in the history of new ideas, from the coffeehouses of eighteenth-century England where citizens gathered to discuss chemistry and radical politics, to the Left Bank bars of modernist Paris frequented by Picasso and Gertrude Stein. The virtue of these third places, Oldenburg says, is that they bring together a diversity of talent, allowing people to freely interact while ingesting some caffeine or alcohol. What makes the Pixar studios so unique is that these spaces have become part of the office itself. There are cubicles and desktops, of course, but there are also whiskey lounges and espresso bars. The end result is a workplace filled with the clutter of human voices, the soundtrack of an effective third place.

....

1. *I eavesdropped on two animators talking about the dirt on Lotso the Bear's fur while washing their hands at the sink.*

While such interactions might seem incidental and inefficient — the kind of casual encounters that detract from productivity — Pixar takes them extremely seriously. The studio knows that the small talk of employees isn't a waste of time, and that those random conversations are a constant source of good ideas. This is because Pixar has internalized one of the most important lessons of group creativity, which is that the most innovative teams are a mixture of the familiar and the unexpected, just like those Broadway artists making *West Side Story*. (The company 3M and Google both promote a similar ethos by emphasizing horizontal interactions.) Although most people at Pixar work in tight-knit teams, the culture of the studio encourages them to chat with colleagues working on completely unrelated projects. "We think a lot about the geography of where people are sitting and how the offices are laid out," Anderson says. "Part of my job [as a producer] is to make sure everyone is smooshing together. If I don't see lots of smooshing, I get worried." If Anderson knows that an animator will be working on a technical aspect of the film, she'll place him at the end of a corridor filled with computer scientists. If a writer is struggling with a scene involving a certain character, then Anderson will make sure the writer bumps into the animators drawing that same character. "The assumption is that a few of those random talks in the hallway are going to be really useful," she says. "Most of them won't be, of course. They'll just be talking about their kids or football or whatever. But every once in a while that random conversation is going to lead to a breakthrough."

In order to understand the wisdom of Pixar's office design, it helps to know about the research of Tom Allen, a professor of organization studies at MIT. In the early seventies, Allen began studying the interaction of engineers in several large corporate laboratories. After several years of tracking their conversa-

THE POWER OF Q

tions—counting all those exchanges in the hallways and coffee room—he came up with the Allen curve, which describes the likelihood that any two people in the same office will communicate. The curve is steep; according to Allen, a person is ten times more likely to communicate with a colleague who sits at a neighboring desk than with someone who sits more than fifty meters away.

It's not particularly surprising, of course, that we make small talk with those who are nearby. But Allen also discovered something unexpected about all these office conversations. After analyzing the workplace data, he realized that the highest-performing employees—those with the most useful new ideas—were the ones who consistently engaged in the most interactions. "High performers consulted with anywhere from four to nine organizational colleagues [on a given project], whereas low performers contacted one or two colleagues at most," Allen wrote in his 1984 treatise *Managing the Flow of Technology*. "This suggests that increasing the number of colleagues with whom an employee consults contributes independently to performance." The key word in that sentence is *independently*. According to Allen's data, office conversations are so powerful that simply increasing their quantity can dramatically increase creative production; people have more new ideas when they talk with more people. This suggests that the most important place in every office is not the boardroom, or the lab, or the library. It's the coffee machine.

A similar lesson emerges from a recent study led by Brian Uzzi, the sociologist who measured the Q of Broadway musicals. In 2009, Uzzi got access to a vast trove of data from a large hedge fund, giving him a complete record of every instant message sent by every trader over an eighteen-month period. The first thing Uzzi and his collaborators discovered was that these traders sent out an astonishing number of messages. They amassed more than two million exchanges, with the average trader engaging in sixteen different IM conversations at the same time.

What Uzzi wanted to know was how this constant stream of information affected the financial performance of the traders. He was particularly interested in the flurry of messages that occurred whenever a new financial report was released. "What you often see is [that] some new information comes over the Bloomberg terminal, and there's this sudden spike in communication," Uzzi says. "What's happening is that everyone is trying to figure out what the news means. Is it good news? What's it going to do to the stock?" Uzzi refers to this as the disambiguation process, since the traders are trying to make sense of the unclear information. They're asking one another questions and benefiting from the diverse thoughts of colleagues. "There's a big incentive to figure this stuff out fast," he says. "The faster you are, the more money you make."

By comparing the messaging habits of the traders, Uzzi was able to document the power of these electronic interactions. He discovered that Tom Allen was right: the best traders were the most connected, and people who carried on more IM conversations and sent more messages also made more money. (While typical traders generated profits on only 55 percent of their trades, those who were extremely plugged in profited on more than 70 percent of their stock trades.) "These are the guys who get embedded in multiple chats," Uzzi says. "They're just sucking up information from everybody else, like a vacuum. And when they start to trade, they don't go silent. They don't stop talking. *They IM even more.*" In contrast, the least successful traders tended to engage in the fewest electronic chats. They also stopped exchanging messages right before making an investment decision. Says Uzzi, "They'd get cut off from the conversation. And this meant that they weren't able to make sense of what was happening."

Uzzi compares these financial conversations to the creative process. "The act of investing is like solving a difficult puzzle," he says. "These traders are trying to connect the dots. And if you look at these IM exchanges [in the hedge fund], what you frequently find is that they lead to a good idea, a successful trade. Because the traders are listening to their network, they manage to accomplish what they could never have done by themselves." While all the instant messages might seem like a distraction—a classic example of multitasking run amok—Uzzi argues that they're an essential element of success. "If I had to choose between a trader who was a little smarter or one who was a little better connected, I'd definitely go for the connections," Uzzi says. "Those conversations count for a lot."

Pixar tries to maximize such conversations. The studio knows that an office in which everyone is interacting is the most effective at generating new ideas, as people chat at the bathroom sink and exchange theories while waiting in line for lattes. "The secret of Pixar from the start has been its emphasis on teamwork, this belief that you can learn a lot from your coworkers," says Alvy Ray Smith. "Ed and I were really determined to create a kind of mutual admiration society, so that the techies thought the artists were geniuses, and the artists thought the techies were magicians. We wanted people to *want* to learn from each other. That's always when the best stuff happens: when someone tells you something you didn't already know."

The computer scientist Christopher Langton once observed that innovative systems constantly veer toward the "edge of chaos," to those environments that are neither fully predictable nor fully anarchic. We need structure or everything falls apart.[2] But we also need spaces that surprise us. Because it is the exchanges we don't expect, with the people we just met, that will change the way we think about everything.

3.

Every day at the Pixar studio begins the same way: A few dozen animators and computer scientists gather in a small screening room filled with comfy velour couches. They eat Lucky Charms and Cap'n Crunch and drink organic coffee. Then the team begins analyzing the few seconds of film produced the day before, ruthlessly shredding each frame. (There are twenty-four frames per second.) No detail is too small to tear apart: I sat in on a meeting in which the *Toy Story 3* team spent thirty minutes discussing the reflective properties of the plastic lights underneath the wings of Buzz Lightyear. After that, an editor criticized the precise starting point of a Randy Newman song. The music began when Woody entered the scene, but he argued that it should start a few seconds later, when Woody began running. Someone else disagreed, and a lively debate ensued. Both alternatives were

2. *The classic demonstration of a space without structure is the "nonterritorial office" developed by the ad agency TBWA Chiat/Day. In 1993, the company decided to do away with every tradition of the corporate office. Employees were no longer given fixed desks or cubicles or computers. Instead, they were encouraged to assemble with their colleagues based on the task at hand; the model was the college campus, in which students were free to work anywhere and everywhere. (Time magazine hailed the Chiat/Day office as the "forerunner of employment in the information age.") The reality of the new space, however, failed to live up to the hype. Although the freeform interior was designed to encourage interaction, it actually derailed it. Because there were no offices, people couldn't find one another. Productivity plummeted. The couches with the nicest views became the subject of petty turf wars; fistfights broke out over meeting spaces. By 1995, it had become clear that the new Chiat/Day model was broken. The walls were reinstalled.*

tested. (It's not uncommon for a Pixar scene to go through more than three hundred iterations.) The team discussed the motivations of the character and the emotional connotations of the clarinet solo. By the time the meeting was over, it was almost lunch.

These crit sessions are modeled on the early production meetings at Lucasfilm, when Alvy Ray Smith, John Lasseter, and Ed Catmull would meet with the animators to review their work. At first, the meetings were necessary because nobody knew what he was doing—computer animation remained a hypothetical. But Lasseter soon realized that the meetings were incredibly efficient, since everybody was able to learn from the mistakes of everybody else. Furthermore, the crit sessions distributed responsibility across the entire group, so that the entire team felt responsible for catching mistakes. "This was a lesson I took away from the Toyota manufacturing process," Catmull says. "In their car factories, everybody had a duty to find errors. Even the lowly guys on the assembly line could pull the red cord and stop the line if they saw a problem. It wasn't just the job of the guys in charge. *It was a group process.* And so what happened at Toyota was a massive amount of incremental improvement. People on the assembly line constantly suggested lots of little fixes, and all those little fixes had a way of adding up to a quality product. That model was very influential for me as we set about figuring out how to structure the Pixar meetings. I wanted people to know that if a mistake slips through the production process, if we don't fix something that can be fixed, then it's everybody's fault. We all screwed up. We all failed to pull the red cord."

The harsh atmosphere of Pixar's morning meetings—the emphasis on finding imperfections and mistakes—may at first seem to contradict one of the basic rules of group creativity, which is to always be positive. In the late 1940s, Alex Osborn, a founding partner of the advertising firm BBDO, came up with a catchy term for what he considered the ideal form of group creativity: *brainstorming*. In a series of best-selling books, Osborn outlined the basic principles of a successful brainstorming session, which he said could double the creative output of a group. The most important principle, he said, was the absence of criticism. According to Osborn, if people were worried about negative feedback, if they were concerned that their new ideas might get ridiculed by the group, then the brainstorming process would fail. "Creativity is so delicate a flower that praise tends to make it bloom, while discouragement often nips it in the bud," Osborn wrote in *Your Creative Power.* "In order to increase our imaginative potential, we should focus only on quantity. Quality will come later."

Brainstorming is the most popular creativity technique of all time. It's used in advertising offices and design firms, the classroom and the boardroom. When people want to extract the best ideas from a group, they obey Osborn's instructions; criticism is censored, and the most "freewheeling" associations are encouraged. The underlying assumption is simple: if people are scared of saying the wrong things, they'll end up saying nothing at all.

There is, of course, something very appealing about brainstorming. It's always nice to be saturated in positive feedback, which is why most participants leave a brainstorming session proud of their contributions to the group. The whiteboard has been filled with free associations, the output of the unchained imagination. At such moments, brainstorming can seem like an ideal mental technique, a feel-good way to boost productivity.

There's just one problem with brainstorming: it doesn't work. Keith Sawyer, a psychologist at Washington University, summarizes the science: "Decades of research have consistently shown that brainstorming groups think of far fewer ideas than the same number of people who work alone and later pool their ideas." In fact, the very first empirical test of Osborn's technique, which was performed at Yale in 1958, soundly refuted the premise. The experiment was simple: Forty-eight male undergraduates were divided into twelve groups and given a series of creative puzzles. The groups were instructed to carefully follow Osborn's brainstorming guidelines. As a control sample, forty-eight students working by themselves were each given the same puzzles. The results were a sobering refutation of brainstorming. Not only did the solo students come up with twice as many solutions as the brainstorming groups but their solutions were deemed more "feasible" and "effective" by a panel of judges. In other words, brainstorming didn't unleash the potential of the group. Instead, the technique suppressed it, making each individual less creative.

The reason brainstorming is so ineffective returns us to the importance of criticism and debate, the very elements that define the Pixar morning meeting. (Steve Jobs has implemented a similar approach at Apple. Jonathan Ives, the lead designer at the company, describes the tenor of group meetings as "brutally critical.") The only way to maximize group creativity—to make the whole more than the sum of its parts—is to encourage a candid discussion of mistakes. In part, this is because the acceptance of error reduces its cost. When you believe that your flaws will be quickly corrected by the group, you're less worried about perfecting your contribution, which leads to a more candid conversation. We can only get it right when we talk about what we got wrong.

Consider this clever study led by Charlan Nemeth, a psychologist at UC-Berkeley. She divided 265 female undergraduates into five-person teams. Every team was given the same difficult problem: How can traffic congestion be reduced in the San Francisco Bay Area? The teams had twenty minutes to invent as many solutions as possible.

At this point, each of the teams was randomly assigned to one of three different conditions. In the minimal condition, the teams received no further instructions; they were free to work together however they wanted. In the brainstorming condition, the teams got the standard brainstorming guidelines, which emphasized the importance of refraining from criticism. Finally, there was the debate condition, in which the teams were given the following instructions:

"Most research and advice suggest that the best way to come up with good solutions is to come up with many solutions. Freewheeling is welcome; don't be afraid to say anything that comes to mind. However, in addition, most studies suggest that you should debate and even criticize each other's ideas."

Which teams did the best? The results weren't even close: while the brainstorming groups slightly outperformed the groups given no instructions, people in the debate condition were far more creative. On average, they generated nearly 25 percent more ideas. The most telling part of the study, however, came after the groups had been disbanded. That's when researchers asked each of the subjects if he or she had any more ideas about traffic that had been triggered by the earlier conversation. While people in the minimal and brainstorming conditions produced, on average, two additional ideas, those in the debate condition produced more than seven. Nemeth summarizes her results: "While the instruction 'Do not criticize' is often cited as the [most] important instruction in brainstorming, this appears to be a counterproductive strategy. Our findings show that debate and criticism do not in-

hibit ideas but, rather, stimulate them relative to every other condition."

There is something counterintuitive about this research. We
naturally assume, like Osborn, that negative feedback stifles the
sensitive imagination. But it turns out we're tougher than we
thought. The imagination is not meek—it doesn't wilt in the face
of conflict. Instead, it is drawn out, pulled from its usual hiding
place.

According to Nemeth, the reason criticism leads to more new
ideas is that it encourages us to fully engage with the work of others. We think about their concepts because we want to improve
them; it's the imperfection that leads us to really listen.[3] (And
isn't that the point of a group? If we're not here to make one another better, then why are we here?) In contrast, when everybody
is "right"—when all new ideas are equally useful, as in a brainstorming session—we stay within ourselves. There is no incentive to think about someone else's thoughts or embrace unfamiliar
possibilities. And so the problem remains impossible. The absence
of criticism has kept us all in the same place.[4]

....

3. *Just look at the Beatles: Lennon and McCartney had a famously combative and competitive relationship. But that turned out to be a blessing in disguise, since all the internal disagreements inspired the songwriters.*

4. *The emotion of anger also seems to have short-term creative benefits. That, at least, is
the take-away message of a 2011 series of studies led by Matthijs Baas, Carsten De Dreu,
and Bernard Nijstad. In their first experiment, they demonstrated that anger was better
than a neutral mood for promoting creativity. In their second experiment, they elicited
anger directly in some of the subjects, and then asked all of the study participants to
brainstorm on ways to improve the environment. Once again, people who felt angry generated more ideas than nonangry people. These ideas were also deemed more original, as
they were thought of by less than 1 percent of the subjects.*

*Of course, this doesn't mean that anger is a cure-all or that nastiness is always wise.
For one thing, anger is resource depleting: although angry subjects generated more ideas
initially, their performance quickly declined.*

To better understand the power of criticism—why it acts like a multiplier for the imagination—it's worth looking at another experiment led by Nemeth. While the typical brainstorming session begins with an instruction to free-associate—to express the very first thoughts that enter the mind—that's probably an ineffective strategy. In study after study, psychologists have found that people just aren't very good at free-associating. For instance, if I ask you to free-associate on the word *blue*, there's a 45 percent chance that your first answer will be *sky*. Your next answer will probably be *ocean*, followed by *green*, and, if you're feeling creative, a noun like *jeans*. Our associations are shaped by language, and language is full of clichés.

How do we escape these clichés? Nemeth found a simple fix. Her experiment went like this: A lab assistant surreptitiously sat in on a group of subjects being shown a variety of color slides. The subjects were asked to identify each of the colors. Most of the slides were obvious, and the group quickly settled into a tedious routine. However, in some groups, Nemeth instructed her lab assistant to occasionally shout out the *wrong* answer, so that a red slide would trigger a response of "Pink," or a blue slide would lead to a reply of "Turquoise." After a few minutes, the group was asked to free-associate on these same colors. The results were impressive: people in the dissent condition—they were exposed to inaccurate descriptions—came up with far more original and varied associations. Instead of saying that *blue* reminded them of *sky*, they were able to expand their loom of associations, so that the color triggered thoughts of *Miles Davis, Smurfs,* and *berry pie*. The obvious answer had stopped being the only answer. More recently, Nemeth has found that the same strategy can lead to improved problem solving on a variety of creative tasks. It doesn't matter if you're trying to invent a new brand name or decipher a hard insight puzzle. Beginning a group session with a moment of dissent—*even when the dissent is wrong*—can dramatically expand creative potential.

The power of dissent is really about the power of surprise. After hearing someone shout out an errant answer—*red* is called *pink*—you start to reassess your initial assumptions. You try to understand the strange reply, which leads you to think about the problem from a new perspective. And so your comfortable associations get left behind. The imagination has been stretched by an encounter that you didn't expect.

These experiments demonstrate the value of Pixar's morning production meetings. When the animators and engineers sit down on those couches with their cereal bowls, they know the meeting isn't going to be very much fun. "Nobody likes to begin their day by learning about all the stuff they got wrong the day before," says Bobby Podesta, the lead animator on *Toy Story 3*. "But we know that, if you want to make the best stuff, then you're going to have to accept some tradeoffs. You're going to have to stay late at the office. You're going to have to deal with critiques. Your feelings might occasionally get hurt."

Nevertheless, Pixar strives to ensure that the criticism never gets out of control, that all the mistakes don't become too demoralizing. This is why the team leaders at Pixar emphasize the importance of plussing, a technique that allows people to improve ideas without using harsh or judgmental language. The goal of plussing is simple: whenever work is criticized, the criticism should contain a plus, a new idea that builds on the flaws in a productive manner. "Since we spend most of our day in these group meetings, it's really important that the meetings stay relatively cordial," Podesta says. "It could get pretty depressing if all we did was shoot each other down. And that's why, when we do engage in criticism, we try to make sure the criticism is mixed with a little something else, a new idea that allows us to immediately move on, to start focusing not on the mistake but on how to fix it."

When plussing works, it's incredibly effective at generating creative breakthroughs. The criticism feels like a surprise, and that makes everyone in the room more likely to invent a plus, a new idea that moves the movie forward. According to Podesta, many of his best fixes come *after* the meeting, as he continues to contemplate the morning conversation. "It might be hours later, but I'm often still thinking about what the group talked about," he says. "Maybe I'm still a little upset because I got taken apart. Or maybe we just exposed a really tough problem, and none of the proposed fixes really worked. But it's like I put the problem on the back burner of my brain. And then, when I'm doing something else"—Podesta can often be found at the Pixar gym—"I come up with a better solution. I suddenly know how I should animate the face, or how that scene should go. I'm still plussing."

This is why the Pixar process is so effective: while the groups engage in critical debate, it is a debate shot through with the unexpected, with the innovative ideas that emerge from relentless dissent. "The most wonderful part of working here are the surprises," says Lee Unkrich, a Pixar director. "Before we begin every movie, there's always the worry that maybe we don't have any good ideas left. Maybe all our good jokes have been used up. But then the process begins and those worries mostly disappear. The team finds a way to make it happen. Because if it was just me making this"—he points to a computer screen with a frame from *Toy Story 3*—"then the movie would stink. I'm not capable of surprising myself every day with some great new idea. That kind of magic can only come from the group."

Sometimes, the dramatic improvements unleashed by the Pixar process can startle outsiders. In August of 2002, Michael Eisner, the CEO of Disney, was given an advance screening of *Finding Nemo,* Pixar's third full-length release. At the time, Disney wasn't sure if it would renew its distribution contract with the fledgling studio. Eisner was not impressed by the film. As James Stewart recounts in *DisneyWar,* the CEO immediately e-mailed the Disney board: "Yesterday we saw for the second time the new Pixar movie *Finding Nemo.* This will be a reality check for those guys. It's OK, but nowhere near as good as their previous films." Eisner used the mediocrity of the movie to explain why he wanted to wait until after its release before restarting contract negotiations with Pixar. The creative failure would allow Disney to get a better deal.

But Eisner was wrong: *Finding Nemo* turned out to be a huge box-office success, grossing more than $868 million. While the rough cut was deeply imperfect, Eisner underestimated the power of Pixar's iterative method. He didn't realize that the studio excelled at fixing its failures, transforming a problematic draft into a polished final cut. (The director Andrew Stanton ended up restructuring the entire movie, cutting a series of flashbacks.) Ed Catmull summarizes this creative journey in typically blunt terms, describing it as the ability to go from "suck to non-suck." The original *Finding Nemo* sucked. But then, after nine months of morning crit sessions, it ended up firmly in the non-suck category, winning the 2003 Academy Award for best animated film. Disney ended up paying dearly for the negotiating delay.

It's important not to sugarcoat the struggles of the Pixar process. Even plussing can't prevent the occasional heated argument, and many employees complain about the grueling hours. ("At least they give us free food on the weekend," Podesta says.) When I spent time at the studio, people answered many of my questions with references to the same traumatic experience: the making of *Toy Story 2*. Although the movie is more than a decade old, it remains a frequently cited parable at Pixar. Catmull, for instance, referred to the struggle of *Toy Story 2* as "our defining moment . . . A lesson we should never forget."

The problems with the film began in the fall of 1998, during the final days of story development.[5] Pixar takes its stories very seriously. In fact, it often takes the studio longer to develop the narrative than to animate the movie. The process begins when the Pixar brain trust—a group composed of John Lasseter, Ed Catmull, and eight directors—hashes out the initial plot, often while sitting at a burger joint down the street. That sketch of a story is then turned into a treatment, a two-page document outlining the basic arc of the movie. Several drafts and plenty of crit sessions later, the treatment is handed over to a screenwriter. (Pixar frequently brings in outside talent to write the scripts. It's one of the many ways they inject fresh voices into the process, ensuring the team maintains the right level of Q.) The studio doesn't want a polished screenplay—it just wants something to get the process started. And so the script gets revised. And then revised again. Scenes are cut; scenes are added. New characters emerge to fill narrative holes. After a year of edits, the script is turned into a story reel, an elaborate sequence of storyboards. There is no ani-

....

5. *Disney originally urged the studio to make the sequel a direct-to-video release, which meant it would have a smaller budget and shorter running time. However, Catmull and Lasseter concluded that the decision was a mistake. "We came to believe that having two different standards of quality was bad for our souls," Catmull says. "You either always make the best stuff you can or you shut up shop."*

mation yet, just drawn poses like in a comic book, with the lines read by Pixar employees. "The reels look very rough," Catmull says. "But they're an essential part of the iterative process. When you see the script as a movie, you see all the mistakes in the story. And there are always many, many mistakes."

It's at this point that *Toy Story 2* began running into serious setbacks. Because the studio had been frantically trying to finish *A Bug's Life,* its second feature film, *Toy Story 2* hadn't benefited from the usual process of plussing. Instead of interacting with the entire studio, the creative team had been largely isolated in a separate building. (The current campus was still under construction.) "The movie was going off course in a way that we had gone off course on the other movies," remembers Unkrich. "But the problem was, we were all so busy trying to get *A Bug's Life* made that we couldn't take the time to help them fix the film, to add our critical voices to the mix."

It wasn't until the winter of 1998 that the brain trust was finally able to start focusing on the troubled cartoon. The first screening of the story reels went horribly. "Everybody knew that the movie wasn't working," says Catmull. "Our process was broken—the story wasn't getting better." And so, with less than a year until the release date, the Pixar team decided to do the unthinkable: they threw the script in the trash and started over. Tom Schumacher, an executive at Disney, was terrified. He remembers the first meeting after the screening:

> John and I were sitting at the table with some of my Disney colleagues, who said, "Well, it's okay." And I can't imagine anything being more crushing to John Lasseter than the expression, "Well, it's okay." It's just unacceptable to him, and it's one of his most endearing, most exasperating qualities, and probably

the biggest reason for his success. So nine months before it was supposed to come out, John threw out the vast majority of the movie. Which is *unheard of.*

How did Pixar fix *Toy Story 2*? The first change was physical. Lasseter immediately moved everyone into the same space, so the engineers and storytellers and directors were all crammed into a small cluster of cubicles. He realized that the movie was missing that Pixar spark, those minor epiphanies and surprising ideas that occur when people interact in unexpected ways. "We decided that from then on we always wanted everybody in one building," Lasseter says. "We wanted all the departments, no matter what movie they were working on, to be together."

Lasseter then scheduled an emergency story summit in Sonoma, a two-day retreat that would give people the freedom to think about the movie in an entirely new way. (The new location turned the team into temporary outsiders.) The brain trust soon realized that the fundamental problem with *Toy Story 2*—the reason the reels weren't working—was that the plot felt too predictable. Although the story revolved around Woody's capture by a toy collector who plans on selling him to a museum in Japan, this scenario never felt like a real possibility. "This film is coming out of Disney and Pixar," Catmull says. "So you already know Woody's going back to his original family in the end. And if you know the end, there's no suspense." Once this narrative flaw was identified, the Pixar team began fixing it. Wheezy, the broken squeaky toy, was moved to the beginning of the film; a plot twist involving the two Buzzes was dramatically expanded; "Jessie's Song," a sad lament about no longer being loved by a child, was inserted into the second act. This intense creative process took its toll, with many team members suffering from stress-related health problems. In *To Infinity and Beyond,* Pixar's official history, Steve Jobs remembers the difficult first months of 1999: "We killed ourselves to make it [*Toy Story 2*]. It took some people a year to recover. It was tough—it was too tough, but we did it."

Toy Story 2 wasn't just finished on time; it went on to become one of the most successful animated films ever made. (The reviews were literally all positive.[6]) Nevertheless, the agonizing production process remains an essential lesson for everyone at the studio. "I'll worry about Pixar when we unlearn what we learned from *Toy Story 2,*" Catmull says. "Meltdowns are always painful, but they're a sign that we're still trying to do something difficult, that we're still taking risks and willing to correct our mistakes. We have to be willing to throw our scripts in the trash." Because Pixar knows that talent is not enough. Talent fails every day. And that's why Jobs put the bathrooms in the center of the building and why the production team begins every day with a group critique. It's why the producers think about where people sit and why the best ideas come when a story is being plussed apart. Everybody at Pixar knows that there will be many failures along the way. The long days will be filled with difficult conversations and disorienting surprises and late-night arguments. But no one ever said making a good movie was easy. "If it feels easy, then you're doing it wrong," Unkrich says. "We know that screwups are an essential part of what we do here. That's why our goal is simple: We just want to screw up as quickly as possible. We want to fail fast. And then we want to fix it. Together."

....

6. *According to Rottentomatoes.com,* Toy Story 2 *is one of the best-reviewed movies of all time, with 146 positive reviews and 0 negative reviews.*

4.

Dan Wieden is cofounder of the advertising agency Wieden+ Kennedy, one of the most innovative and honored ad agencies in the world. Wieden's firm has a reputation for designing unconventional campaigns, from the Levi's commercial featuring the voice of Walt Whitman to those yellow rubber bracelets that support Lance Armstrong's foundation. The agency created the classic Michael Jordan Nike ads and produced a Miller beer television spot directed by Errol Morris. Its employees conceived of the viral Old Spice ads on YouTube and reinvented *SportsCenter* with the satirical "This is *SportsCenter*" campaign.

I met Wieden at the W+K headquarters in the Pearl District of Portland, Oregon. The building is a former cold-storage factory that's been hollowed out. This means that the interior is mostly empty space, a soaring lobby framed by thick concrete walls and weathered pine beams. Wieden gives me a tour of the building as he explains his unorthodox approach to fostering group creativity.

At first glance, the Wieden+Kennedy office can seem like a case study in creativity lite, dense with the kind of "innovation enhancers" that fill the pages of business magazines. There's modern art on the walls[7] and the coffee room is plastered with invitations to team-building exercises, including pie-making competitions and company-sponsored trips to the museum. While Dan believes in the virtue of such events—he's particularly proud of the biannual pub-crawl—he thinks they work only if the right people are present. For Dan, this is what creativity is all about: putting tal-

---

7. *The office feels like a gallery; every surface is covered with art. My favorite installation is a huge white canvas filled with tens of thousands of clear plastic pushpins. It's only when you take a step back that the mural makes sense. The pushpins spell the following slogan: Fail Harder.*

ented people in a room and letting them freely interact. "It really is that simple," he says. "You need to hire the best folks and then get out of the way."

How does Wieden find these people? How does he ensure that his office is filled with employees who will inspire one another? Wieden takes the problem of hiring so seriously that, in 2004, he decided to start his own advertising school, which he called WK12. (The name is a misnomer, since the school actually consists of thirteen people who work together for thirteen months.) There are no classes at WK12. Instead, the curriculum consists of real assignments from real clients, with the students working under the direction of seasoned Wieden+Kennedy employees. The advantage of the school, Wieden says, is that it allows him to not worry about experience—"CVs can be so misleading"—and instead focus on those intangible qualities that are essential for creativity. "What I've learned to look for is the individual voice," he says. "It might be an aesthetic, or a sentence style, or a way of holding the camera. But having that unique voice is the one thing I can't teach. I can teach someone to write copy. I can show someone how to crop a photo. But I can't teach you how to have a voice. You either have something to say or you don't."

Not surprisingly, the applicants to WK12 come from every conceivable field. A recent graduating class included a struggling poet, a grad student in anthropology, a chemist, a chef, a cinematographer, and two novelists. (The advertisements for WK12 feature a single question: "Tired of a pointless life?") For Wieden, the school is an important means of ushering in fresh blood, forcing the agency to incorporate new voices from new disciplines. The inexperienced students ask naive questions and come up with plenty of impractical suggestions. They turn in assignments late and can't figure out the technical equipment. "You could look at these students, and you could easily conclude that they are wasting everyone's time," Wieden says. "They don't know what the hell they're doing."

But that's the point. Wieden describes the challenge of advertising as finding a way to stay original in a world of clichés, avoiding the bikinis in beer ads and the racing coupes in car commercials. And that's why he's so insistent on hiring people who don't know anything about advertising. "You need those weird fucks," he says. "You need people who won't make the same boring, predictable mistakes as the rest of us. And then, when those weirdos learn how things work and become a little less weird, then you need a new class of weird fucks. Of course, you also need some people who know what they're doing. But if you're in the creative business, then you have to be willing to tolerate a certain level of, you know, weirdness." Wieden is describing the advertising version of Brian Uzzi's research on Broadway musicals, as the constant influx of students ensures that his creative teams remain in the sweet spot of Q. And so, every year, a new class of WK12 students walks into the headquarters of Wieden+Kennedy and sets up shop in the lobby. Most of their work will be thrown away. Most of their drafts will be ignored. But their weirdness will be contagious.[8]

One of Wieden's favorite stories illustrates the importance of incorporating a little weirdness into the creative process. In 1988, Wieden was hard at work on a series of television spots for Nike. The campaign consisted of eight video clips, each of which fo-

----

8. David Ogilvy, one of the founding fathers of modern advertising, pursued a similar approach. When Ogilvy tested his ideas for a particular marketing campaign, he always included several pitches that he was sure would not work. "Most were, as expected, dismal failures," Ogilvy wrote. "But the few that succeeded pointed to innovative approaches in the fickle world of advertising."

cused on a different athlete in a different sport. Wieden knew that the campaign needed a tag line, a slogan that could link the disparate commercials together. Unfortunately, he was drawing a blank. "I'd been struggling to find that line for months," he says. "And it was late at night, and we had to have it ready to go in the morning. And so I'm getting nervous, thinking about how this really wouldn't work without a slogan. But I couldn't come up with a slogan! It was killing me."

But then, just when Wieden was about to give up and go to sleep, he started thinking about a murderer named Gary Gilmore who had been executed in 1977. "He just popped into my mind," Wieden says. "And so it's the middle of the night, and I'm sitting at my desk, and I'm thinking about how Gilmore died. This was in Utah, and they dragged Gilmore out in front of the firing squad. Before they put the hood over his head, the chaplain asks Gilmore if he has any last words. And he pauses and he says: 'Let's do it.' And I remember thinking, 'That is so fucking courageous.' Here's this guy calling for his own death. And then, the next thing I know, I'm thinking about my shoe commercials. And so I start playing around with the words, and I realized that I didn't like the way it was said, actually, so I made it a little different. I wrote 'Just Do It' on a piece of paper and as soon as I saw it, I knew. That was my slogan."

The question, of course, is why Wieden started thinking about Gary Gilmore while working on a slogan for cross-trainers. "I swear, I don't normally think about murderers at midnight," he says. "So I asked myself: Where did this thought come from? And the only explanation I could come up with is that someone else in the group"—one of his colleagues working on the Nike campaign—"had mentioned Norman Mailer to me earlier in the day. I don't know why Mailer came up. I can't remember. I'm sure we were just bullshitting, doing what people do when you put them in a small room together. But we were talking about Mailer, and I knew that he'd written a book about Gary Gilmore. And that was it. That's where the slogan came from. Just a little sentence from someone else. That's all it takes."

7 URBAN FRICTION

*By its nature, the metropolis provides what otherwise
could be given only by traveling; namely, the strange.*
—Jane Jacobs

DAVID BYRNE LOVES bicycles. He's been relying on bikes to
get around since the late 1970s, when he started riding a beat-
up three-speed all over Manhattan. At the time, the bicycle was
mostly a convenience, an easy way to escape the downtown traffic. "I could run errands in the daytime and efficiently hit a few
clubs, art openings, or nightspots in the evening without searching for a cab," Byrne says. "The bike wasn't cool, and I almost got
killed a few times, but it was better than driving." In the years
since, Byrne has become a self-described bicycle fanatic. He now
takes a folding bike wherever he goes; he's pedaled to rock concerts in Dallas and researched an opera by riding in the streets of
the Philippines. He's gotten lost in the ghetto of Detroit and cy-
cled through the hectic alleys of Istanbul.

While Byrne celebrates the pleasures of biking—"The wind
in your face, the exercise, the relaxation"—he bikes mostly for another reason: *it lets him listen to the city.* He describes cycling as

a form of urban eavesdropping, a way to overhear the hum of the streets. "When you're stuck in a car, it's like you're in a bubble," Byrne says. "You can't hear anything that's happening outside. But when you're on a bike, you can tap into the atmosphere. You can feel people doing their thing. It's a kind of connection."

When I met Byrne outside his office loft in SoHo, on a cobblestone street filled with fancy clothing boutiques, he was carrying a muffin and a helmet; his shock of white hair was perfectly vertical. He led me inside, up three flights of stairs, and down a grim, industrial hallway. (The building used to be a sweatshop.) It was a warm day, and the windows of his studio were wide open — the sound of the street seeped in. "I like it a little noisy," Byrne says. "It reminds me where I am."

David Byrne is a rock 'n' roll legend. For sixteen years, he was the lead singer of Talking Heads, the new wave band that invented the new wave. (In 2002, the group was inducted into the Rock and Roll Hall of Fame.) But Byrne isn't just a singer of the classic pop songs "Burning Down the House," "Heaven," and "Once in a Lifetime"; he's also an avant-garde composer, visual artist, and bicycle activist. In 1981, he pioneered the use of sampling with Brian Eno on their album *My Life in the Bush of Ghosts.* (The layered, cacophonous sound of the album would later influence the development of hip-hop.) Recently, Byrne wrote a disco opera about Imelda Marcos, designed bike racks for New York City, and retrofitted an old ferry terminal so that the metal pipes could be played like a church organ. "I don't worry very much about how all of my projects go together," Byrne says. "I don't think about how this fits with that or what came before. I just get an idea and then I follow it."

Where do all of Byrne's ideas come from? His answer is simple: the city. It is the muse that inspires his music, the noisy source of his art. It's why Byrne bicycles to work from Hell's Kitchen and keeps the windows of his office wide open. He first discovered the creative potential of the city after dropping out of the Rhode Island School of Design to start a punk band. Byrne moved to Manhattan to be close to every other punk band, the place where groups like Television and Blondie were redefining the aesthetics of rock. And so Talking Heads—this scraggly group of design students—began playing small clubs in the East Village. (One of their first gigs was opening for the Ramones at CBGB.) "We started with really small audiences, maybe twenty people," Byrne says. "We'd only make money when they'd buy beer." Because the clubs were mostly empty, Byrne and the band were able to hone their craft, experimenting with their sound. "Nobody comes out of their basement playing perfect," Byrne says. "Most of the time, you don't even know what you want to play. And that's why it was so important for us to have these places that were a little forgiving."

After a late show, Byrne would often bike around the neighborhood to unwind. Sometimes he'd venture over to a stretch of Latin dance clubs by the East River. "I was the only white guy there, but I would just hang out, enjoying these intense rhythms," he says. "It was all new to me." Around the same time, Byrne was introduced to the music of Fela Kuti. He quickly became obsessed. ("The grooves were so intense, trance-inducing almost," Byrne remembers. "I couldn't help but want to steal that sound.") And then there was the art scene. When Byrne wasn't hanging around ethnic clubs, he was staring at a Jasper Johns painting in a SoHo gallery or admiring one of Andy Warhol's soup-can prints. (Warhol was an early fan of Talking Heads.) This urban amalgam—the mixture of ethnic sounds and new ideas percolating in the downtown streets—profoundly shaped Byrne's early rock 'n' roll compositions. He describes the process as mostly involuntary: "If you look and listen in a city, then your mind gets expanded automatically," Byrne says. "You can say 'I know that's possible because I saw somebody else do this.' And then you take that and maybe without even knowing it you start to put it in your own music."

This is what makes Talking Heads such an important band: they were one of the first groups to fuse these diverse influences, to make music that blended melodies and ideas from all over the world.[1] It was punk music distorted by the polyglot sounds of New York City, the beat of the disco as interpreted by the postmodern-art scene. The end result was a new way of thinking about what sounds could exist together. "The city definitely made it possible," Byrne says. "A lot of what's in the music is stuff that I first heard because it was playing down the street . . . Those are the accidents that have always been so important for me. And they just happen naturally in the right place."

For Byrne, the metropolis is like a sonic blender; every street is a mix tape. Cities expand the imagination by exposing us to unexpected things, to funky Latin beats and jangly Nigerian bass lines and abstract works of art. And then, when we're in the studio, we can't help but weave these ideas into our own work, so that punk rock is melded with pop paintings, Afro-Cuban rhythms, and symbolist choreography. This is why Byrne describes cities as a kind of "energy source," and why he always bikes with a dicta-

----

1. *A list of singers and bands influenced by Talking Heads includes everyone from Peter Gabriel to Paul Simon to Vampire Weekend.*

phone in his pocket. "You never know when an idea is going to come to you," he says. "Cities are not just about all the cultural stuff. That's nice, but that's not it. In a vibrant city, you can get just as much from going to the barbershop, or walking down a crowded street, as you can from going to a museum. It's about paying attention and listening to everything that's happening. It's about letting all that stuff in, so the city can change you."

1.

Why do cities exist? This is a surprisingly difficult question to answer. The modern metropolis, after all, can be an unpleasant, expensive, and dangerous place. It's full of rush-hour traffic and panhandlers, overpriced apartments and feisty cockroaches. The air is dirty, there is litter in the streets, and the public schools are falling apart. In other words, urban life isn't easy. We cram ourselves together, but all the cramming comes with a cost.

Thomas Malthus, the eighteenth-century British economist, was the first to focus on the cost. In his 1798 *Essay on the Principle of Population* — the work would later influence Charles Darwin — Malthus argued that a surfeit of people inevitably led to a shortage of things. This led Malthus to conclude that cities were doomed and that their steady expansion — London grew from 1 million people in 1800 to 6.7 million a century later — would lead to a future of "extermination, sickly seasons, epidemics . . . and gigantic inevitable famine." Human density was always undone by earthly scarcity.

But Malthus was mostly wrong; London today is not full of starving citizens. While the pessimistic economist assumed that cities were a doomed social experiment, his forecast was exactly backward. Instead of dying out, cities have boomed; urban growth is the great theme of modern life, a migratory trend that's unfolding all across the world, from the factory boomtowns of southern China to the sprawling favelas of Rio de Janeiro. In fact, for the first time in history, the majority of human beings live in urban areas. (The numbers are far higher in developed countries; the United States, for instance, is 81 percent urbanized.) In the next century, more people will move to cities than have moved to cities in all of human history.

What explains the rapid urbanization of the world? What was Malthus's mistake? In an influential 1988 paper, the Nobel Prize–winning economist Robert Lucas concluded that the continued vitality of cities was a fundamental paradox. According to "the usual list of economic forces," Lucas wrote, "the city should fly apart . . . The theory of production contains nothing to hold a city together." Although economists could quantify the burdens of urban life—those pricey condos and violent crimes—they struggled to understand the advantages. It remained a mystery why all these strangers were squishing themselves together.

While Lucas didn't have any good answers—the economic equations, he said, were entirely useless—he did endorse the speculations of Jane Jacobs, an urban activist and author of *The Death and Life of Great American Cities*. Jacobs first got interested in cities as a way of defending Greenwich Village, her neighborhood. At the time, these small-scale enclaves were under constant attack as city planners sought to "modernize" the civic landscape, bulldozing old buildings and erecting "super-blocks" filled with residential high-rises and elevated highways. It was Tomorrowland today; just as technology had revolutionized the private spaces of the home with the introduction of gizmos such as the dishwasher and television, so science would transform our public spaces. Urban blight would soon be a thing of the past.

But Jacobs wasn't convinced. She begins *The Death and Life* by describing the failure of the first wave of urban renewal:

> Look what we have built with the first several billions [in redevelopment spending]: Low-income centers that become worse centers of delinquency, vandalism and general social hopelessness than the slums they were supposed to replace . . . Cultural centers that are unable to support a good bookstore. Civic centers that are avoided by everyone but bums, who have fewer choices of loitering place than others . . . Promenades that go from no place to nowhere and have no promenaders. Expressways that eviscerate great cities. This is not the rebuilding of cities. This is the sacking of cities.

The debacle of modern urban planning led Jacobs to explore the virtues of old-fashioned neighborhoods. She began by stepping out her front door and analyzing a stretch of Hudson Street in the Village. Jacobs compared the crowded sidewalk to a spontaneous "ballet," filled with people from different walks of life. There were school kids on the stoops, and gossiping homemakers, and "business lunchers" on their way back to the office. There was Mr. Lacey the locksmith chatting with Mr. Slube at the cigar store, and the Irish longshoremen drinking beer next to the poets at the White Horse Tavern. While urban planners had long derided such neighborhoods for their inefficiencies—that's why Robert Moses, the master builder of New York, wanted to build an eight-lane elevated highway through SoHo and the Village—Jacobs argued that these casual exchanges were essential. She saw the city not as a mass of buildings but as a vessel of empty spaces in which people interacted with other people. The city wasn't a skyline—it was a dance.

Furthermore, these sidewalk conversations came with real benefits. According to Jacobs, the virtue of Hudson Street was that it encouraged the "mingling of diversity," allowing city dwellers to easily exchange information.[2] Although cities might suffer from a shortage of physical resources, Jacobs emphasized their surplus of human capital, which produced valuable innovations like new wave music.

And this is why the Village was so vital. The neighborhood might look like an anachronism—it was designed for horse carts, not cars—but Jacobs insisted that its layout remained the urban ideal. The Village had short city blocks, which were easier for pedestrians to navigate. It had lots of old buildings—Jacobs's street was mostly nineteenth-century tenements and townhouses—with relatively cheap rents, and cheap rents encouraged a diversity of residents. Most important, the streets were mixed use, filled with apartments and retail shops and restaurants, which meant that different kinds of people were on the street for different reasons at different times of the day. The end result was a constant churn of ideas as strangers learned from one another. Jacobs coined a telling phrase for what happens in these densely populated spaces: "knowledge spillovers."

What's interesting is that the sheer disorder of the metropolis maximizes the amount of spillover. Because cities force us to mingle with people of different "social distances"—we have dinner parties with friends, but we also talk to strangers on the street—we end up being exposed to a much wider range of

....

2. She also liked their aged architecture. As Jacobs famously remarked, "New ideas require old buildings." What she meant is that more recent buildings—the planned developments of suburbia, for instance—tend to be too expensive for risky, arty, and small-scale businesses. "Chain stores, chain restaurants and banks go into new construction. But neighborhood bars, foreign restaurants and pawn shops go into older buildings. Supermarkets and shoe stores often go into new buildings; good bookstores and antiques dealers seldom do." In other words, it's not an accident that CBGB and all those other punk clubs sprang up in the oldest part of Manhattan.

worldviews. While it's tempting to discount these urban inter-actions—what could possibly emerge from a random sidewalk chat?—they actually come with impressive payoffs. Look, for in-stance, at a study led by Adam Jaffe, an economist at Brandeis University. He analyzed the paper trail of patent citations, which is the list of previous inventions cited in every patent application. Jaffe found that innovation was largely a local process; citations were nearly ten times as likely to come from the same metropoli-tan area as a control patent. This suggests that inventors are in-spired by other inventors in their neighborhoods, even when the research involves entirely unrelated subjects. And this logic doesn't apply just to patents. David Byrne, after all, wasn't influ-enced by the Latin rhythms of some distant musician. Instead, Byrne was seduced by his local dance clubs blasting those songs he could hear from the sidewalk. It is the sheer density of the city—the proximity of all those overlapping minds—that makes it such an inexhaustible source of creativity.[3]

When I ask Byrne if the city continues to define his art, he responds with a story about his most recent music tour. "A few months before the tour I decided I wanted to have some dancers in the show," he says. "I know a couple of choreographers in the neighborhood, and I ran into one of them, so I asked her if she could suggest anybody. And she gave me a few names, and then those people gave me a few names, and so on. After a few days, I'd found my dancers and they were perfect. Now, I could have done

....

3. *Interestingly, cities and brains seem to have converged on the same solution to the problem of connectedness. The neuroscientist Mark Changizi has demonstrated that ur-ban areas and the human cortex rely on extremely similar structural patterns to maxi-mize the flow of information and traffic through the system. In other words, a neural highway acts just like its concrete counterpart: "When scaling up in size and function, both cities and brains seem to follow similar empirical laws," Changizi said. "They have to efficiently maintain a high level of connectedness in order to work properly."*

it without using my local friends. I could have held an open audition and gone through all that. But that's so much work, I never would have done it. I just wouldn't have had any dancers. And that would have been a shame, because dancers make everything better."

2.

Geoffrey West doesn't eat lunch. His doctor says he has a mild allergy to food; meals make him sleepy and nauseated. When West is working—when he's staring at some scribbled equations on scratch paper or gazing out his window at the high desert—he subsists on caffeinated tea and the occasional sugar cookie. His gray hair is tousled, and his beard has the longish look of neglect. It's clear that West regards the mundane needs of everyday life—feeding the body, trimming the whiskers—as little more than annoying distractions that draw him away from much more interesting problems. Sometimes West can seem jealous of his computer, this silent machine with no hungers or moods. All it needs is a power cord.

For West, the world is most compelling at its most abstract. He likes to compare himself to Kepler, Galileo, and Newton, since he's also a theoretical physicist in search of fundamental laws. But West isn't trying to decode the physical universe; he's not interested in deep space, black holes, or string theory. Although West worked for decades as a physicist at Stanford University and Los Alamos National Laboratory—his specialty was the behavior of elementary particles—he left the field after the Texas superconducting supercollider was canceled by Congress in 1993. "At first, I was devastated," West says. "I had all these great experiments planned." West wasn't ready to retire, however, and so he began thinking about what to study next. "I realized that what physicists are very good at is finding laws," he says. "We're good at making sense of complexity."

And so West began searching for a subject that needed his skill set. He eventually settled on cities. For the physicist, the urban jungle seemed like one of those systems that looked chaotic—all those taxi horns and traffic jams—but might actually be obeying a short list of cosmic rules. "We spend all this time thinking about cities in terms of their local details, in terms of their neighborhoods and restaurants and museums," West says. "I had this hunch that there was something more, that every city was also shaped by a set of hidden laws."

But West realized that no one was looking for these laws. He saw modern urban theory as akin to physics before Kepler; he ridiculed it as a field without principles. "It's all just speculation," West says. "It's storytelling. There is no rigor." If he was going to dispense advice and tell mayors how to run their cities, then he wanted his advice to be wholly empirical, rooted in the strictness of facts. West was tired of urban theory—he wanted to invent urban *science*.

Of course, before West could solve the city—transforming the speculations of Jane Jacobs into a quantitative discipline—he needed data. Massive amounts of data. Along with Luis Bettencourt, another physicist who had given up on physics, West began scouring the libraries for urban statistics. The scientists downloaded huge files from the U.S. Census, learned about the intricacies of German infrastructure, and spent several thousand dollars on a thick almanac featuring the provincial cities of China. (Unfortunately, the book was in Mandarin.) They looked at a dizzying array of variables, from the length of electrical wire in Frankfurt to the number of college graduates in Boise to the average income in Shenzhen. They amassed stats on gas stations and personal income, sewer pipes and murders, coffee shops and the walking speed of pedestrians.

After two years of careful analysis, West and Bettencourt discovered that all of these urban variables could be described by a few exquisitely simple equations. These are the laws that automatically emerge whenever people "agglomerate," cramming themselves into apartment buildings, subway cars, and sidewalks. It doesn't matter if the city is Manhattan, New York, or Manhattan, Kansas; the urban patterns remain the same. West isn't shy about describing the magnitude of this accomplishment. "What we found are the constants that describe every city," West says. "I can take these laws and make very accurate predictions about the number of violent crimes and the surface area of roads and the average income in a city in Japan with two hundred thousand people. I don't know anything about this city. I don't know where it is, or its history, or what it produces, but I can tell you all about it. And the reason I can do that is because every city is really the same. They're all the same damn thing, and that's why the equations work."

The existence of these equations depends on the ballet of Hudson Street. While Jacobs could only speculate on the value of our urban interactions, West insists that his equations confirm her theory. He describes his data as a scientific version of Jacobs's sidewalk dance, since it quantifies the value of urban spaces. "One of my favorite compliments is when people come up to me and say, 'You have done what Jane Jacobs would have done, if only she could do mathematics,'" West says. "What the numbers clearly show, and what she was clever enough to anticipate, is that when people come together they become much more productive per capita. They exchange more ideas and generate more innovations. What's truly amazing is how predictable this is. It happens automatically, in city after city."

According to the equations of West and Bettencourt, every socioeconomic variable that can be measured in cities—from the production of patents to per capita income—scales to an exponent of approximately 1.15. What's interesting here is the size of the exponent, which is greater than 1. This means that a person living in a metropolis of one million should generate, on average, about 15 percent more patents and make 15 percent more money than a person living in a city of five hundred thousand. (The one living in the bigger city should also have 15 percent more restaurants in his neighborhood and create 15 percent more trademarks.) The correlations remain the same even when the numbers are adjusted for levels of education, work experience, and IQ scores. "This remarkable equation is why people move to the big city," West says. "Because you can take the same person, and if you just move them from a city of fifty thousand to a city of six million, then all of a sudden they're going to do three times more of everything we can measure. It doesn't matter where the city is or which cities you're talking about. The law remains the same."

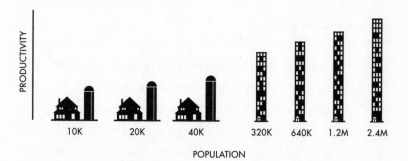

Cities exhibit superlinear growth: as they get bigger, every person in the city becomes more productive.

West and Bettencourt refer to this phenomenon as "superlinear scaling," which is a fancy way of describing the increased output of people living in big cities. When a superlinear equation is graphed, it looks like a roller coaster climbing into the sky. The steep slope emerges from the positive-feedback loop of urban life—a growing city makes everyone in that city more productive, which encourages more people to move to the city, and so on. According to West, these superlinear patterns demonstrate why cities are the single most important invention in human history. They are the idea, he says, that enabled our economic potential and unleashed our ingenuity. Once people started living in dense clumps, they created a kind of settlement capable of reinventing itself, so a city founded on the fur trade could one day give birth to Wall Street, and an island in the Seine chosen for its military advantages might eventually become a place full of avant-garde artists. "Cities are this inexhaustible source of ideas," West says. "And that's entirely because of these equations. As cities get bigger, everything starts *accelerating*. Each individual unit becomes more productive and more innovative. There is no equivalent for this in nature. Cities are a total biological anomaly. But you can't understand modern life without understanding cities. They are the force behind everything interesting. They are where everything new is coming from."

After West and Bettencourt discovered the superlinear laws that define every city, they became interested in what the equations couldn't explain. While the math was able to predict the approximate performance of a given urban area, it failed to describe the local deviations, those slight differences that made Bridgeport feel distinct from Brooklyn, or New Orleans distinct from Seattle, or Austin distinct from Houston. "When I tell people about our urban laws, the first question they always ask is about these differences," West says. "They say: 'If every city is the same, then why does my city seem so unique?' Eventually, we got so sick of this question that we decided to find the answer."

The first thing the physicists discovered is that these deviations persisted over time. Bridgeport, for instance, has been abnormally wealthy for at least a century, while New Orleans has always had an extremely high crime rate, and Austin has consistently produced many more patents per capita than Houston. "That's when I realized it wasn't such a silly question after all," West says. "These deviations aren't just fluky things or random short-term trends. Instead, they seem to reflect something real about the city itself."

Furthermore, the endurance of these differences means that it's possible to uncover the hidden correlations of urban life. What is it about Austin, for instance, that makes it the most innovative city in Texas? Why has Cleveland always been so poor? Why does Santa Monica generate so many trademarks? While West and Bettencourt have discovered many interesting patterns — the best way to reduce crime, for instance, is to attract young college graduates — one of their most surprising findings comes from an old survey led by the Princeton psychologists Marc and Helen Bornstein. In the early 1970s, the Bornsteins began measuring the average walking speed of pedestrians in dozens of different cities all across the world, from Geneva to Jerusalem. The research itself was straightforward: the Bornsteins measured out a sixty-foot stretch of sidewalk and then timed thousands of random pedestrians as they walked down the street. Interestingly, the psychologists discovered that the pace of life closely tracked the population of cities, so the more people in the city, the faster they walked.

While the physicists cite this paper as yet another piece of evidence in support of superlinear scaling—density speeds us up—they've also discovered an unexpected correlation. There seems to be a consistent link between walking speed and the production of patents: cities with unusually fast pedestrians create more new ideas. (Both urban measurements are extremely consistent over time.) West and Bettencourt attempt to explain this peculiar link by invoking the language of physics. They compare urban residents to particles with velocity bouncing off one another and careening in unexpected directions. The most creative cities are simply the ones with the most collisions.

Of course, these interpersonal collisions—the human friction of a crowded space—can also feel unpleasant. We don't always want to talk with strangers on the subway or jostle with other pedestrians. Nevertheless, West insists that all successful cities are a little uncomfortable. He describes the purpose of urban planning as finding a way to minimize people's distress while maximizing their interactions. The residents of Hudson Street, after all, didn't mind talking to one another on the sidewalk or chatting with the butcher while buying their meat. As Jacobs pointed out, the layout of her Manhattan neighborhood—the irregular grid, the density of brownstones—meant that it didn't feel like the center of a vast metropolis that was overstuffed with strangers. It maintained an intimate feel and facilitated the most meaningful kind of mingling. That's why it's called the Village.

In recent decades, however, many of the fastest-growing cities in America, such as Phoenix and Riverside, have pursued a very different urban model. These places have focused on mitigating unwanted interactions, trading away crowded public spaces and knowledge spillovers for single-family homes. However, West and Bettencourt point out that these suburban comforts are closely associated with poor performance on a variety of urban metrics. (Several economic studies have found that doubling urban density raises productivity by up to 28 percent.) Phoenix, for instance, has had below-average levels of income and innovation for the last forty years. "When you look at some of these fast-growing cities, they look like tumors on the landscape," West says with his usual bombast. "They have these extreme levels of growth, but it's not sustainable growth. That's because they haven't developed the necessary kind of interactions that lead to new ideas. You can't grow forever on the promise of cheap land."

West contrasts the long-term underperformance of Phoenix with San Jose, a city that has been an unusually innovative place for the last hundred years. (The data is compelling: while Phoenix was ranked 146th among American cities in the production of patents per capita, San Jose was second.[4]) In fact, even when the San Jose region was mostly walnut and apricot farms, and decades before it became the center of Silicon Valley, the area still produced an abnormally large number of patents per capita. "There is just something about this city that makes it really good at generating patents," West says. "This local culture is why Silicon Valley is in that particular valley. It's not a historical or geographical accident—it's because of these numbers. Now, the equations can't tell you *why* the San Jose area has always been so innovative. All they can show is that Silicon Valley is an unusually creative place. It's a real outlier."

....

4. *According to West and Bettencourt, the most innovative city in America is Corvallis, Oregon. However, this ranking is slightly misleading, since a disproportionate number of the patents generated in Corvallis come from a single Hewlett-Packard lab that conducts research on laser printers.*

3.

Route 128 is a highway around Boston. It begins as a two-lane road in the fishing port of Gloucester, Massachusetts, and then bends inland toward the suburbs. In the early 1950s, the highway became shorthand for the American high-tech industry, which was scattered along its off ramps. (In 1955, *BusinessWeek* referred to Route 128 as "the Magic Semicircle," while *Forbes* called it "America's Technology Highway.") This was particularly true around Waltham and Newton, two towns that were soon populated by industrial parks and glassy office towers. By 1970, the area bounded by Route 128 contained six of the ten largest technology firms in the world, including Digital Equipment Corporation and Raytheon. The "Massachusetts Miracle" was under way.

While Route 128 was undergoing a postwar boom, the San Jose region remained heavily agricultural; the main local industry was small-scale food-processing plants. This is why it was so surprising when, in 1956, William Shockley, the eccentric co-inventor of the transistor, established the Shockley Transistor Corporation in a small town called Mountain View. (Shockley had tried to start a transistor company in the Route 128 region, but the large Boston firms weren't interested in his product.) Because Shockley couldn't convince any of his former colleagues from Bell Labs to join his new venture—nobody wanted to move to a farming town in California—he ended up recruiting grad students from Caltech and Stanford. Unfortunately, Shockley turned out to be a terrible manager, and in 1957, eight of his researchers left the company. This group of refugee engineers—they called themselves the Traitorous Eight—founded the Fairchild Semiconductor Company in a San Jose garage. They sold their very first batch of transistors to IBM, and by 1963, they had sales of more than $130 million a year. A few years later, two of these engineers—Robert Noyce and Gordon Moore—left Fairchild to start their own microchip company. They called it Intel.[5]

By the early 1980s, Silicon Valley was home to dozens of success stories like Intel, including Apple Computer, Cisco, Oracle, and Sun Microsystems. In fact, these young start-ups were so successful that by 1985, Silicon Valley had nearly twice as many people working in high-tech as did the Route 128 area. In the years since, the West Coast advantage has only grown: Internet companies like Netscape, Google, Netflix, and Facebook have all emerged from the suburbs around San Jose. (Although Facebook was founded in a Harvard dorm room in February of 2004, Mark Zuckerberg moved the company to Palo Alto that summer. He said he wanted to be close to the action.) And the original tech stalwarts of the Boston area—those massive companies like Digital Equipment Corporation and Wang Laboratories—have all gone out of business. In fewer than fifty years, the walnut farms of San Jose have become the technology center of the world.

The astonishing success of Silicon Valley raises an interesting question: What happened to Route 128? As Vivek Wadhwa, a business professor at Duke, notes: "If you were betting on an area to dominate [the tech sector] in 1975, you'd have been wise to bet on Route 128. It had a giant head start over everywhere else." The region had several elite research universities, such as MIT and Harvard, and a long list of successful technology firms. These

....

5. *A third member of the Traitorous Eight went on to form the venture-capital giant Kleiner Perkins (now KPCB), which was an early investor in many of the most successful tech companies, including Amazon, AOL, Compaq, EA, Genentech, Google, Netscape, and Sun Microsystems.*

companies had big contracts with the Defense Department and controlled the market for microchips and electronic hardware.

And yet, this head start wasn't enough: after a few decades of domination, the Boston high-tech sector began to fall apart. This decline was primarily rooted in the inability of Route 128 companies to keep pace with the innovations of the San Jose region. AnnaLee Saxenian, a professor of city planning at UC-Berkeley, compares the postwar performance of these two regions in her insightful book *Regional Advantage*. She argues that by the mid-1970s, the Boston area was already "several years behind the curve . . . set by California companies. They couldn't compete with the new designs and products coming out of Silicon Valley."

What caused this innovation gap? The downfall of the Boston tech sector was caused by the very same features that, at least initially, had seemed like such advantages. As Saxenian notes, the Route 128 area had been defined for decades by the presence of a few large firms. (At one point, Digital Equipment Corporation alone employed more than a hundred and twenty thousand people.) These companies were so large, in fact, that they were mostly self-sufficient. Digital Equipment Corporation didn't make just minicomputers—it also made the microchips in its computers, and it designed the software that ran on those microchips. (Gordon Bell, the vice president in charge of research at Digital, described the company as "a large entity that operates as an island in the regional economy.") As a result, the Boston firms took secrecy very seriously—a scientist at Digital wasn't allowed to talk about his work with a scientist at Wang or share notes with someone at Lotus. These companies strictly enforced noncompete clauses and nondisclosure agreements; former employees couldn't work for competitors, and scientists weren't allowed to publish articles in peer-reviewed journals. This meant that at the Route 128 companies, information tended to flow *vertically,* as ideas and innovations were transferred within the firms.

While this vertical system made it easier for Route 128 companies to protect their intellectual property, it also made them far less innovative. This is because the creativity of an urban area depends upon its ability to encourage the free flow of information—we need that knowledge spillover—as all those people in the same Zip Code exchange ideas and work together. But this didn't happen around Route 128. Although the Boston area had a density of talent, the talent couldn't interact—each firm was a private island. The end result was a stifling of innovation.

The vertical culture of the Boston tech sector existed in stark contrast to the horizontal interactions of Silicon Valley. Because the California firms were small and fledgling, they often had to collaborate on projects and share engineers. This led to the formation of cross-cutting relationships, so that it wasn't uncommon for a scientist at Cisco to be friends with someone at Oracle, or for a cofounder of Intel to offer management advice to a young executive at Apple. (We saw earlier how these horizontal interactions can trigger moments of insight.) These networks often led to high employee turnover as people jumped from project to project: in the 1980s, the average tenure at a Silicon Valley company was less than two years. (It also helped that noncompete clauses were almost never enforced in California, thus freeing engineers and executives to quickly reenter the job market and work for competitors.) This meant that the industrial system of the San Jose area wasn't organized around individual firms. Instead, the region was defined by its professional networks, by groups of engineers trading knowledge with one another. Tom Wolfe, in a 1983 profile of Robert Noyce, described the informal "schmoozing" that defined Silicon Valley:

Every year there was some place, the Wagon Wheel, Chez Yvonne, Rickey's, the Roundhouse, where members of this esoteric fraternity, the young men and women of the semiconductor industry, would head after work to have a drink and gossip and brag and trade war stories about phase jitters, phantom circuits, bubble memories, pulse trains, bounceless contacts, burst modes, leapfrog tests, p-n junctions, sleeping-sickness modes, slow-death episodes, RAMs, NAKs, MOSes, PCMs, PROMs, PROM blowers, PROM burners, PROM blasters, and teramagnitudes, meaning multiples of a million millions.

These casual exchanges—the errant conversations that take place in coffee houses and bars—are an essential engine of innovation. While Jane Jacobs might have frowned upon the sprawl of these California suburbs, the engineers have managed to create their own version of Greenwich Village. They don't bump into one another on the crowded sidewalk or gossip on the stoop of a brownstone. Instead, they meet over beers at the Wagon Wheel, or trade secrets at the Roundhouse. It's not the ballet of Hudson Street, but it's still a dance, and it's the dance that matters.

The Homebrew Computer Club is a perfect example of the type of socializing that defines Silicon Valley. Founded in the spring of 1975 in a Menlo Park garage, the club began when a few "microcomputer enthusiasts" placed the following invitation on bulletin boards throughout Silicon Valley: *Are you building your own computer? If so, you might like to come to a gathering of people with like-minded interests. Exchange information, swap ideas, help work on a project, whatever.* Before long, the club outgrew the garage and was moved to an auditorium in the Stanford physics department. Although the meetings were informal, they typically followed the same basic structure. The session would begin with a mapping period that allowed people to shout out questions, speculate on rumors, and share their future plans. Then came a presentation featuring a club member discussing a recent invention. Finally, there was the random-access session, in which everyone wandered around the auditorium, networking with strangers.

This engineering club played an important role in the development of Silicon Valley. According to Steve Wozniak, a cofounder of Apple, the first computers built by the company weren't designed to be profitable consumer products; they were built to show off at the Homebrew Club: "The machines were meant to bring down [to the club] and put on the table," Wozniak wrote in his memoir *iWoz*. "I wanted to impress people. Look at this, it uses very few chips. It's got a video screen. You can type stuff on it. Personal computer keyboards and video screens were not well established then. Schematics of the Apple I were passed around freely, and I'd even go over to people's houses and help them build their own."

The club was defined by these friendly collaborations, by the horizontal interactions of engineers meeting in their spare time. As Wozniak continued to develop the Apple I, he incorporated feedback from other members. They'd tell him about upcoming microprocessors and help troubleshoot his circuit board. They'd give him advice on working with floppy-disk drives and offer suggestions on negotiating with suppliers. According to Wozniak, the innovations of the first Apple computers depended entirely on this Homebrew culture: "Today it's pretty obvious that if you're going to build a billion-dollar product, you have to keep it secret while it's in development because a million people will try to steal it," he says. "If we'd been intent on starting a company and selling our product, we'd probably have sat down and said, 'Well, we have to choose the right microprocessor, the right number of characters on the screen,' etc. All these decisions were being made by other companies, and our computer would have wound up being like theirs—a big square box with switches and lights, no video terminal built in . . . It would have probably been a big failure."

But Wozniak's machines were different. Instead of designing a machine like those of the Route 128 firms, he transformed the advice of his diverse social network into an inexpensive computer, just a twenty-dollar microprocessor shackled to 256k of memory. It was still a limited device, but it represented an important breakthrough in usability. And then, in the spring of 1976, Wozniak started talking to another member of the Homebrew Club. This engineer was convinced that the Apple kit could be the foundation for a new kind of computer firm, one that would sell its products directly to consumers. Wozniak, however, wasn't interested: he had a solid job at Hewlett-Packard and didn't want to become an entrepreneur. But this pushy engineer refused to take no for an answer. After a few weeks, Wozniak was "worn down" by his arguments. And so, on April 1, 1976, Apple Computer was founded by Steve Wozniak and his friend from the club. His name was Steve Jobs.

4.

It's four in the afternoon and Yossi Vardi is still in his pajamas. We're meeting at his house, located in the Tel Aviv suburbs on a quiet street near the university. It's the day after a national holiday, and it's clear that Vardi has been out late. He apologizes for the yawns and sprawls across his couch, resting his hands on his rotund belly. He looks like he could fall asleep again at any moment.

I'm here to talk to Vardi about the Israeli technology boom, which is unfolding largely in the suburbs of Tel Aviv. The place feels like Silicon Valley must have felt in the late 1970s, over-stuffed with young entrepreneurs toting business plans. On the outskirts of the city, the tomato farms and olive groves are quickly being transformed into generic office parks. Consider the statistics: in 2008, Israel attracted nearly $2 billion in venture-capital (VC) funding; that is, investments made in small companies with high growth potential. (VC funding is widely regarded as one of the best measures of innovation; money chases good ideas.) This means that a country of seven million people attracted as much funding as France and Germany *combined*. It means that Is-rael—a tiny sliver of land on the Mediterranean, about the same size as Vermont—has nearly three times the amount of VC fund-ing per capita as the United States and thirty times the average of Western Europe.

But Israel was not always a tech center dense with successful start-ups. In the 1990s, the country was best known for military hardware and agricultural products; it built advanced radar sys-tems and exported avocados. "There was very little civilian tech investment," Vardi says. "Nobody talked about Israeli software or microchips or batteries. They talked about drip irrigation." And yet, by 2009, Israel had more companies listed on NASDAQ than Canada. Eric Schmidt, the CEO of Google, has said that Israel is the second-best place in the world for entrepreneurs, after the United States. Steve Ballmer, the CEO of Microsoft, has de-clared that "Microsoft is an Israeli company as much as an Ameri-can company," since much of its most important R & D now takes place in Tel Aviv. (The Kinect, for instance, was largely designed by Israeli engineers.) In the last decade, Israel has produced more successful high-tech start-ups than Japan, India, Korea, and the United Kingdom.

Yossi Vardi is at the center of this boom. He has helped fund more than seventy technology companies in Israel and has taken twelve of them public. (Sergey Brin, the cofounder of Google, once said: "If there is an Internet bubble in Israel, then Vardi is the bubble.") Although Vardi has been a fixture in Israeli politics since the mid-1970s, his involvement in the tech sector began in November of 1996, when Vardi and his oldest son, Arik, launched ICQ, the first online chat program for Windows users. "This was all my son's idea," Vardi says. "I barely even knew what the Internet was." Nevertheless, Vardi saw the potential in live chat and decided to fund the start-up. "So we release this program in November, and it starts to spread right away," he remembers. "After a few months we had a hundred thousand members. And I thought that was amazing. Here was this brand-new tool, with a hundred thousand people using it! But then it just kept on going. By July [1997], we had one million users. September, two million. December, six million. I'd never seen anything like this. And it wouldn't stop. When we got bigger than the AOL chat program, AOL bought us." The price tag was more than $400 million.

In recent years, Vardi has taken a leading role in building the Israeli tech sector, transforming an initial burst of enthusiasm—the sale of ICQ made the front page of every Israeli newspaper—into a genuine Silicon Valley rival. He has started cell phone companies, search engines, social networking sites, and solar-power plants. And yet, as Vardi notes, this success has been achieved despite seemingly impossible odds. Just think of the obstacles: Israel exists on an arid stretch of land surrounded on all sides by hostile neighbors. The nation spends approximately 10 percent of its GDP on defense, which is the highest level among developed nations. It has few natural resources, lurches from war to war, and has absorbed more than 1.5 million immigrants in the last twenty years.

But Israel has turned all of these drawbacks into assets. Consider the small size of the country in terms of both its land and population. "The tiny scale of this country is extremely important," Vardi says. "In Israel, the social graph is very simple: everybody knows everybody. And if you don't know somebody, then they are probably only one degree removed. You can find them very easily." (It also helps that Israel is the second-densest country in the developed world, with more than 91 percent of people living in urban areas.) The intimacy of Israeli social networks means that ideas circulate at an incredibly fast pace; the knowledge spillovers are nonstop. Just look at Vardi, who insists that he learns about all of his start-ups from casual conversations with other people. "What typically happens is a friend tells me about his friend, who has this interesting idea," he says. "And so then I talk to some other people, who also think the idea is interesting. And maybe they talk to some other people, and before you know it we've got a funding plan. That's how the process always works."

Vardi is referring to a very particular kind of social connection. It's long been clear that the vast majority of people have somewhere between four and seven close friends, or what sociologists refer to as strong ties. This number is remarkably stable across all cultures and demographics, suggesting that we're inherently limited when it comes to cultivating deep relationships. After all, there's only so much time in the day for telephone chats and get-togethers.

However, the same uniformity doesn't apply to weak ties or

those people seen only on occasion. The number of weak ties varies dramatically from person to person and can be deeply influenced by culture and place. Some people, like Yossi Vardi, cultivate a vast network of acquaintances; at one point, I watched as Vardi interrupted a cell phone conversation to take a call on his landline and then put both on hold to send a quick e-mail. He attends nearly fifty conferences every year on an array of different subjects and rarely turns down a social invitation. "Most of my day is about making new connections," Vardi says. "The more the better."

At first, it's easy to dismiss such casual relationships. Why should the number of acquaintances matter? These are people we rarely see; their influence seems insignificant. Nevertheless, weak ties turn out to be an essential ingredient of creativity, which is why those cities that encourage an expanded social circle, such as Tel Aviv and San Jose, are more innovative. Martin Ruef, a sociologist at Princeton, has investigated the importance of these contacts for individual entrepreneurs. He began by interviewing 766 graduates of the Stanford Business School, all of whom had gone on to start their own businesses. Ruef was most interested in the structure of their social networks. He noticed that most entrepreneurs had a relatively limited circle of contacts. They might have plenty of good friends, but all of their friends came from the same place and were interested in the same things. Instead of forming weak ties with people at different companies, they invested in relationships with people who were close by. This isn't particularly surprising: We all naturally self-segregate, choosing to spend time with people who are just like ourselves.[6]

6. *Sociologists refer to this failing as the self-similarity principle. In 2007, Paul Ingram and Michael Morris conducted a study of business executives at Columbia University. The executives were invited to a cocktail mixer, where they were encouraged to network*

But not every entrepreneur had such a confining social network. In fact, Ruef discovered a small subset of businesspeople who cultivated a large number of weak ties and unexpected friendships. Instead of hanging out with colleagues and close friends, these entrepreneurs had social networks that were expansive and diverse, full of surprising interactions and "informational entropy." (A system has entropy when it's defined by the presence of disorder—think of a crowded sidewalk.) These businesspeople chatted with acquaintances at conferences and struck up conversations with random strangers at their local coffee places. In other words, they lived like Yossi Vardi, constantly exposed to a wide range of people.

Ruef then analyzed each of these entrepreneurs using an elaborate metric of innovation. He measured the number of patents they'd applied for and kept track of all their trademarks. He rated the originality of their products and gave them bonus points if they'd "entered an unexploited niche" or pioneered a new marketing method. He then compared these innovation rankings to the structure of each entrepreneur's social network. The results were astonishing: businesspeople with entropic networks full of weak ties were *three times* more innovative than people with small networks of close friends. Instead of getting stuck in the rut of confor-

....

with new people. Not surprisingly, the vast majority of executives at the event said their primary goal was to meet "as many different people as possible" and "expand their social network." Unfortunately, that's not what happened. By surreptitiously monitoring the participants with electronic devices, Ingram and Morris were able to track every conversation. What they found was that people tended to interact with others who were most like them, so that investment bankers chatted with other investment bankers, and marketers talked with other marketers, and accountants interacted with other accountants. Instead of interacting with new people, the students at the mixer made small talk with those from similar backgrounds; the smallness of their social world got reinforced. According to Ingram and Morris, the only successful networker at the event was the bartender.

mity—thinking the same tired thoughts as everyone else—they were able to invent profitable new concepts.

There is something unsettling about Ruef's data. We think of entrepreneurs, after all, as creative *individuals*. If someone has a brilliant idea for a new company, we assume that he or she is inherently more creative than the rest of us. This is why we idolize people like Bill Gates and Richard Branson and Oprah Winfrey. But Ruef's analysis suggests that this focus on the singular misses the real story of innovation. The most creative ideas, it turns out, don't occur when we're alone. Rather, they emerge from our social circles, from collections of acquaintances who inspire novel thoughts. Sometimes the most important people in life are the people we barely know.

The unexpected strength of weak ties also helps to explain the Israeli tech boom. As Vardi notes, the connectedness of the country accelerates the pace of innovation: when strangers trade knowledge, new knowledge is created. For instance, after ICQ was purchased by AOL, there was an influx of young graduates into the tech space; everybody was chasing the same lucrative dream. "All of a sudden, it seemed like every kid with even the faintest idea for an Internet company was starting one," Vardi says. "And even though most of these new ideas weren't very good, and most of these companies failed very quickly, the competition was great for everybody. There was this big surge of new people, which allowed us to build a high-tech community in a very short period of time. And that meant we could all start learning from each other." This was the real virtue of ICQ: it transformed a small clump of strong ties into a sprawling network of weak ones.

Of course, Israel isn't the only place trying to attract venture-capital investment. Cities have been attempting to replicate the

Silicon Valley model for decades, striving (often in vain) to create the kind of region from which companies like Google and Apple emerged. So how did Israel—this tiny nation with a long list
of disadvantages—become such a vibrant high-tech incubator?
What is it about the weak ties of Tel Aviv that make its entrepreneurs such a consistent source of new ideas?

One of Vardi's answers is surprising: the military. Because Israel is an extremely small country, it's never been able to maintain
a large standing army. As a result, the Israel Defense Forces (IDF)
are entirely dependent on their reserve units; most Israeli men
under the age of forty-five are required to serve several weeks a
year.

While many Israelis complain about the mandatory service—
it's long been seen as a drag on economic productivity—there's
a growing recognition that it's also a crucial source of innovation.
The reason, Vardi says, is that the reserve forces help maintain a
vast network of weak ties across the country, since the soldiers re-
acquaint themselves with everyone else in their unit every year.
Instead of just hanging out with close friends, they are forced to
mingle with all sorts of different people. "It's like a college re-
union that goes on for a month," Vardi says. "You meet people,
you schmooze. Reserve duty is a big part of what helps keep this
place feeling so tiny." This socializing helps explain how a single
unit in the Israel Defense Forces could give rise to three successful tech companies: ICQ, Check Point Software, and NICE
Systems.

The larger point is that Israel's social networks have come to
resemble those in the most effective urban neighborhoods. What
Jane Jacobs celebrated about Hudson Street is also true of Tel
Aviv. These places have found a way to maximize their human
capital, forcing their inhabitants to mingle.

5.

In the late 1990s, when the dot-com fever was at its peak, many technology enthusiasts predicted that cities would soon become obsolete, a relic of the analog age. After all, with an online world of e-mail and video chats, why should we sacrifice our quality of life to live amid strangers? Cheap bandwidth would mean the end of expensive rents; the zeros and ones hurtling across the fiber-optic cables would supply each of us with all of our human interactions.

And yet this pessimistic forecast for cities has not come to pass. In fact, the data suggests that the opposite has occurred: cities have become more valuable than ever.[7] Edward Glaeser, an economist at Harvard, has studied the effects of the Internet on face-to-face exchanges. Interestingly, he's found that the online world has *increased* the returns of such conversations, at least as measured by metrics such as rents in urban areas and attendance at industry conferences. "Modern life has made being smart and creative even more important," Glaeser says. "And how do we get smart? Even in this age of technology, we still get smart being around other smart people. That's why we pay a fortune to live in Manhattan or Cambridge or Silicon Valley. The printing press didn't make cities less important. And the Internet won't either."

Take a study led by researchers at the University of Michigan. They brought several groups of people together and had them play a difficult cooperation game. Then they gave the exact

---

7. *Just look at the continued vitality of Silicon Valley. One might expect high-tech companies to embrace the possibility of remote communication, but this hasn't happened. Instead, the tech sector continues to concentrate itself in a single California valley, despite the high rents. That's because the best innovators in the world know that the best way to interact is face to face.*

same task to a different set of groups, except these people had to communicate electronically, using e-mail and instant messaging. The groups meeting in person quickly solved the problem, finding clever ways to cooperate. The electronic groups, in contrast, struggled to interact. Although they exchanged roughly equivalent amounts of information, these groups were missing the surprising exchanges that occur when people meet in the flesh. As a result, their problem remained impossible.

A similar lesson emerges from a 2010 study by Isaac Kohane, a researcher at Harvard Medical School. Kohane's question was simple: How does physical proximity affect the quality of scientific research? To find out, he analyzed more than thirty-five thousand peer-reviewed papers, mapping the precise location of every single coauthor. (According to Kohane, this task took a "small army of undergraduates" several years to complete.) Once the data was amassed, the correlation became clear: when coauthors were located closer together, their papers tended to be of significantly higher quality, as measured by the number of subsequent citations. In fact, the best research was consistently produced when scientists were working within ten meters of one another, while the worst papers tended to emerge from collaborators located a kilometer or more apart. "If you want people to work together effectively, these findings reinforce the need to create architectures and facilities that support frequent physical interactions," Kohane says. In other words, our most important new ideas don't arrive on a screen. Rather, they emerge from idle conversation, from too many scientists sharing the same space.

This doesn't mean we should give up on the Internet. Instead, the limitations of technology should inspire us to rethink the nature of our online interactions. The first thing we have to ensure is that our new digital contacts don't detract from our real connections, from the analog conversations of the physical world. Twitter has its uses, but it's no substitute for the ballet of Hudson Street, just as the weak ties of Facebook cannot replace the weak ties of life. While the Internet has evolved primarily to maximize efficiency and make it as easy as possible to find information, it needs to do a better job of increasing *serendipity*. Sometimes the most important idea is the one we don't even know we need.

And this returns us to the enduring power of the metropolis, a place that constantly introduces us to the unexpected and curious. When we stroll down a crowded sidewalk, we meet people we didn't want to meet and get answers to questions we didn't even think to ask; knowledge leaks from everywhere. If the Internet is going to become an accelerator of creativity, then we need to design websites that act like our most innovative cities. Instead of sharing links with just our friends, or commenting anonymously on blogs, or filtering the world with algorithms to fit our interests, we must engage with strangers and strange ideas. The Internet has such creative potential; it's so ripe with weirdness and originality, so full of people eager to share their work and ideas. What we need now is a virtual world that brings us together for real.

The importance of cities is a testament to the necessity of sharing ideas. It doesn't matter if this sharing takes place on Hudson Street or at a bar full of engineers or during army reserve training—the exchange is all that matters. What's interesting is that this urban dance cannot be choreographed in advance or controlled from above. Instead, the creativity of the metropolis is inseparable from its freedom, from the natural chaos of a densely populated Zip Code.

Geoffrey West makes this clear by comparing cities to corporations. At first, urban areas and companies look very similar. Each is a large agglomeration of people interacting in a well-defined physical space. They both contain infrastructure and human capital; the mayor of the city is like the CEO of the corporation.

But it turns out that cities and companies differ in one very fundamental regard: cities almost never die, while companies are extremely ephemeral. As West notes, a cataclysmic hurricane couldn't wipe out New Orleans, and a massive nuclear bomb failed to erase Hiroshima from the map. In contrast, the modern corporation has an average lifespan of only forty-five years. This fragility doesn't apply to just small companies; only two of the original twelve companies in the Dow Jones Index are still in business, while 20 percent of the companies listed in the Fortune 500 disappear every decade.

This raises the obvious question: Why are corporations so fleeting? After spending twenty-five thousand dollars for statistics on more than eighty-five hundred publicly traded companies, West and Bettencourt discovered that corporate productivity, unlike urban productivity, didn't increase with size. In fact, the opposite happened: as the number of employees grew, the amount of profit per employee *shrank*. (While cities are superlinear, companies are sublinear, scaling to an exponent around 0.9.) According to West, this decrease in per capita production is rooted in a failure of innovation. Instead of imitating the freewheeling city, these businesses minimize the very interactions that lead to new ideas. They erect walls and establish hierarchies. They keep people from relaxing and having insights. They stifle conversations, discourage dissent, and suffocate social networks. Rather than maximizing

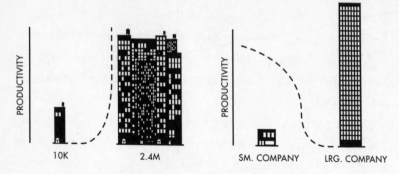

While the per capita creativity of cities rapidly increases with size, companies exhibit the opposite trend.

employee creativity, they become obsessed with minor efficiencies.

The danger, of course, is that this shortage of useful new ideas eventually leads to a decline in profits, which makes a large company increasingly vulnerable to market volatility. Since the company now has to support an expensive staff—overhead costs increase with size—even a minor disturbance can lead to massive losses, because the company is unable to adapt. "The psychology of Wall Street is that companies can never stop growing," West says. "But the sublinear nature of the data suggests that such growth comes with real disadvantages."

For West, the impermanence of the corporation illuminates the real strength of the metropolis. Unlike companies, which are managed in a top-down fashion by a team of highly paid executives, cities are unruly places, largely immune to the desires of politicians and planners. "Think about how powerless a mayor is," West says. "Mayors can't tell people where to live or what to do or who to talk to. Cities can't be managed, and that's what keeps them so vibrant. They're just these insane masses of people bumping into each other and maybe sharing an idea or two. And it's that spontaneous mixing, all those unpredictable encounters, that keeps the city alive."

West illustrates the same point when talking about the Santa Fe Institute (SFI), the think tank where he and Bettencourt work. The institute itself is a disjointed collection of common areas, old couches, and tiny offices; the coffee room is always the most crowded place. "SFI is all about the chance encounters," West says. "There are no planned meetings, just lots of unplanned conversations. It's like a little city that way." (It's also a bit like Pixar—everybody has to use the same bathrooms.) When I was visiting the institute, West and I ran into the novelist Cormac McCarthy in the lobby of the building, where McCarthy does much of his writing. The physicist and the novelist ended up talking about fish without gills, the editing process, and convergent evolution for forty-five minutes. "It's moments like that that make this place so great," West says before listing all the recent ideas to come out of SFI.[8] "It might seem like we're just bullshitting here, wasting time. And I guess maybe we are. But that's also where all the breakthroughs come from."

This is the purpose of cities: The crowded spaces force us to interact. They lead us to explore ideas that we wouldn't explore on our own, and converse with strangers we'd otherwise ignore. The process isn't always pleasant—there's a reason people move to the suburbs—but it remains essential. The superlinear equations of West and Bettencourt quantify this remarkable process, meas-

----

8. *A short list of the important ideas pioneered at SFI includes emergence, biological scaling, chaos theory, and quantum cosmology.*

uring the predictable surges in innovation that happen whenever people share the sidewalk with lots of other people. Sometimes these encounters will lead a person to invent a new patent or think about an old problem in a slightly different way. And sometimes they'll lead him down the street to a Latin dance club where he'll hear a rhythm he's never heard before. It is the human friction that creates the sparks.

8 THE SHAKESPEARE PARADOX

No man is an island.
—John Donne

A FEW YEARS ago, David Banks, a statistician at Duke University, wrote a short paper called "The Problem of Excess Genius." The problem itself is simple: human geniuses aren't scattered randomly across time and space. Instead, they tend to arrive in tight, local clusters. (As Banks put it, genius "clots inhomogeneously.") In his paper, Banks gives the example of Athens between 440 B.C. and 380 B.C. He notes that the ancient city over that time period was home to an astonishing number of geniuses, including Plato, Socrates, Pericles, Thucydides, Herodotus, Euripides, Sophocles, Aeschylus, Aristophanes, and Xenophon. These thinkers essentially invented Western civilization, and yet they all lived in the same place at the same time. Or look at Florence between 1450 and 1490. In those few decades, a city of less than fifty thousand people gave rise to a staggering number of immortal artists, including Michelangelo, Leonardo da Vinci, Ghiberti, Botticelli, and Donatello.

What causes such outpourings of creativity?[1] Banks quickly dismissed the usual historical explanations, such as the importance of peace and prosperity, noting that Athens was engaged in a vicious war with Sparta, and that Florence had recently lost half its population to the Black Plague. He also rejected "the paradigm thing" hypothesis, which argues that genius flourishes in the wake of a major intellectual revolution. The problem with this explanation, Banks said, is that it fails to account for all the paradigm shifts that did not inspire a burst of brilliance. The academic paper concludes on a somber note. "The problem of excess genius is one of the most important questions I can imagine, but very little progress has been made," Banks wrote. The phenomenon remains a mystery.

And yet it's not a total mystery. While we may never know how Athens gave rise to Plato or why Florence became such a center of artistic talent, we can begin to make sense of the clustering of geniuses. The excess is not an accident.

1.

When William Shakespeare arrived in London, sometime in the mid-1580s, the city was in the midst of a theatrical boom. There were more than a dozen new playhouses, many of which staged a different play six days a week. On a typical night, approximately 2 percent of Londoners went to see a performance, with more than a third attending at least one play a month. This meant that the theater industry was both extremely competitive—there were at least a dozen different companies—and hungry for new talent. And so, although Shakespeare had little theatrical experience,

---

1. *The musician Brian Eno has a clever name for these local bursts of innovation. They are examples, he says, of "scenius," which is the "communal form of genius."*

he left behind his wife and two young children and moved to London.

Shakespeare's new hometown was one of the densest settlements in human history. Approximately two hundred thousand people were packed into a few square miles on the banks of the Thames. In fact, the demand for space was so high that the newest neighborhoods were stacked on top of old graveyards; basement walls were full of bones. While this unprecedented density came with drawbacks—riots and plague were a constant threat—it also had its economic advantages: wages in the metropolis were about 50 percent higher than elsewhere in the country. (It turns out that the superlinear equations of West and Bettencourt applied even in the sixteenth century.) As a result, London continued to attract throngs of young people like Shakespeare. It's estimated that by 1590, more than half of the city's population was under the age of twenty.

The playhouses were at the center of this human maelstrom; they were the densest places in the densest city. Most of these new theaters were built on the outskirts of London, next to the brothels, prisons, and lunatic asylums. Land was cheaper here, but the playhouses also benefited from being just beyond the city line, which meant they were able to operate largely without regulation. Shakespeare probably performed for the first time at the Rose, a brand-new playhouse in Southwark with plaster walls and a thatched-straw roof. Although the inner yard of the Rose was only forty-six feet in diameter, it could accommodate an audience of nearly two thousand, giving it a density three times that of the typical modern playhouse.

In 1587, shortly after Shakespeare arrived in London, the Rose introduced *Tamburlaine the Great,* a new play by Christopher Marlowe. It was an epic drama, full of rampaging chariots, live cannons, and fake blood. While the plot was fairly straightfor-ward—it's about a Scythian shepherd who rises to become a dom-inant king—Marlowe pioneered a new kind of dramatic writing known as blank verse, in which the lines are bound by their meter and not their rhymes. This new literary form allowed Marlowe to fill the play with natural-sounding speech. It moved theater away from poetry—Marlowe castigated "rhyming mother-wits"—and toward narratives driven by their characters.

For the impressionable Shakespeare, *Tamburlaine* was a rev-elation. Marlowe had shown him what was possible onstage, creat-ing a work that was both popular and profound. He had bent the predictable arc of the morality play into something more interest-ing, a vernacular drama with eternal themes. This was mass en-tertainment that did more than entertain. This play lingered in the mind.

But *Tamburlaine* must also have been difficult for Shake-speare to watch. Surely it was tough to see a playwright with such a similar biography—Marlowe had also been born in 1564 to a commoner father in a provincial town—create the kind of play that he himself wanted to create. Furthermore, Marlowe had been blessed with a crucial privilege that Shakespeare had been denied: a university education. In the early 1580s, Marlowe had been awarded a full scholarship to the University of Cambridge. (The scholarship was intended to encourage "poor students of promise . . . such as can make verse.") Marlowe decided to major in the arts and spent much of his time translating Ovid and Virgil.

Shakespeare would have witnessed the benefits of this edu-cation in *Tamburlaine*. While Marlowe borrowed the plot from a series of popular history books, he enriched the play with exotic details of Persia and Turkey; the script was full of knowing refer-ences to harems and banquets and old African kings. One of Marlowe's main sources for these facts was an extremely expensive manuscript by Abraham Ortelius, a Flemish geographer. Marlowe couldn't afford the book, but that didn't matter: Cambridge had two copies.

How could Shakespeare compete with Marlowe? He didn't have access to a vast academic library or the prestige that came with a university education. (In 1592, the playwright Robert Greene castigated Shakespeare for being an uneducated "upstart crow.") Instead of translating ancient Roman poetry at Cambridge, Shakespeare had spent his formative years as an actor, learning the craft from the inside. He seemed destined to perform the lines of other men; he would never write anything that would last.

But Shakespeare was saved by his time. Thanks to a mixture of new institutions and policies, Elizabethan England proved to be the ideal place for a young dramatist to develop. It was, for one thing, an age obsessed with the theater: no society had ever been so eager to see plays performed on the stage. As a result, novice writers like Shakespeare were able to get work and gain experience.

And yet even as Londoners flocked to the playhouse, they were also pioneering a new kind of literary culture in which books became an important part of the public discourse. This is largely because sixteenth-century England underwent a massive increase in literacy—there hadn't been this many readers in a city since ancient Athens. While historians estimate that less than 1 percent of English citizens could read in 1510, by the time Shakespeare moved to the capital, the literacy rate was approaching 50 percent.

This was not the case in other countries. A government survey of Catholic France, for instance, estimated that for the period from 1686 to 1690, more than 75 percent of the population could not sign their names. One explanation for these differences in literacy is the emphasis on textual interpretation in Protestant countries like England. Because the Reformation encouraged the masses to read, it also helped create a new market for books—more than seven thousand titles were published during the reign of Elizabeth. (By 1600, the neighborhood around Covent Garden contained nearly one hundred independent publishers; residents complained that the streets reeked of ink.) The end result was a dramatic democratization of knowledge, and Shakespeare and his contemporaries gained access to a vast number of new stories and old texts. These playwrights didn't need Cambridge or Oxford—they had the bookstore.

The surplus of literature proved especially important in the beginning of Shakespeare's career. In 1587—the very same year that Shakespeare began writing his first play—Richard Fields, a childhood friend of his from Stratford who had become a printer, published Raphael Holinshed's *Chronicles of England, Scotland and Ireland*. Shakespeare couldn't afford the book, but it seems likely that Fields allowed the young playwright to treat his store like a lending library. Before long, Shakespeare was mining the *Chronicles* for plots, searching for stories that could compete with the spectacle of *Tamburlaine*. He began with the reign of Henry VI, an era when the English court was torn apart by petty jealousy. There was lust and blood, love and vengeance, murder and conspiracy. Shakespeare had found his subject.

The play premiered at the Rose Theater in 1592, and it was an instant theatrical success. The London crowds were thrilled to watch their own history onstage; there was something electrifying about seeing an English king played by a commoner. And then there was the violence. Shakespeare had learned from Marlowe that audiences relished blood, and so he filled the scenes with sword fights and stab wounds. By the time the historical trilogy was complete, Shakespeare had become one of the most popular playwrights in London.

Shakespeare, however, was only beginning to learn his craft. In fact, the Henry VI plays were so deeply influenced by *Tamburlaine* that eighteenth-century scholars assumed Marlowe had written most of the lines. It didn't help that several of the characters were crudely drawn caricatures utterly lacking the humanity of Shakespeare's later creations. If these plays were all that was known of Shakespeare, then today we wouldn't know him. His writing wouldn't deserve to be remembered. It's not that Shakespeare wasn't a gifted young playwright—it's that he had yet to become a genius.

And this is why culture matters. While Shakespeare is often regarded as an inexplicable talent—a man whose work exists outside of history—he turns out to have been profoundly dependent on the age in which he lived. It was the welter of Elizabethan England that inspired him to become a playwright and then allowed him to transform himself from a poor imitation of Marlowe into the greatest writer of all time. Shakespeare is a reminder, in other words, that culture largely determines creative output.

Unfortunately, this is often because our culture holds us back. Instead of expanding the collective imagination, we make it harder for artists and inventors to create new things. We stifle innovation and discourage the avant-garde. We get in the way of our geniuses.

Every once in a while, however, we get it right. We stumble upon an ideal cultural mix that allows people to create in new ways. The result is a surplus of geniuses, an outpouring of talent so extreme we assume it can never happen again.

And this brings us back to Elizabethan England. Here is a society that got it right, allowing its writers to reach their full creative potential. Consider the list of geniuses who surrounded Shakespeare. There was Marlowe, of course, but also Ben Jonson, John Milton, Sir Walter Raleigh, John Fletcher, Edmund Spenser, Thomas Kyd, Philip Sidney, Thomas Nash, John Donne, and Francis Bacon. Although most of these men came from modest backgrounds, they were able to invent a new kind of literature. We are still scribbling in their shadows.

The triumph of Shakespeare is intertwined with this literary flourishing. Just look at his library, which must have been filled with a staggeringly diverse collection of texts. If Coleridge was, as they say, the last man to have read everything, then Shakespeare was the first.[2] While his shelves featured respectable fiction like the *Tragicall Historye of Romeus and Juliet,* the *Decameron,* and the *Heptameron,* they also included a wide range of popular romance stories. (Shakespeare had a particular weakness for Italian pulp fiction.) He drew from Ovid and Plutarch, but he also borrowed from history books, giving his kings lines straight from the *Chronicles.* And then there were the popular pamphlets, the literature of the street, which Shakespeare constantly worked into his plays. Shakespeare read Edmund Spenser and Chaucer—his English predecessors—but he also read younger poets, like John Donne, in manuscript form. And then there were Shakespeare's fellow playwrights, whom he tracked like a spy. He studied Thomas Watson's sonnets and ripped off the story of Greene's *Pandosto.* (In fact, virtually all of Shakespeare's plots, from *Hamlet* to *Romeo and Juliet,* were adapted from other sources.) He never stopped stealing from Marlowe.

----

2. *Prospero, in* The Tempest: *"My library / Was dukedom large enough."*

But Shakespeare didn't just read these texts and imitate their best parts; he made them his own, seamlessly blending them together in his plays. Sometimes this literary approach got Shakespeare into trouble. His peers repeatedly accused him of plagiarism, and he was often guilty, at least by contemporary standards. What these allegations failed to take into account, however, was that Shakespeare was pioneering a new creative method in which every conceivable source informed his art. For Shakespeare, the act of creation was inseparable from the act of connection.

Ben Jonson famously wrote that Shakespeare "was not of an age, but for all time!" On the one hand, Jonson was right: *Macbeth* and *King Lear* and *The Tempest* are immortal works of art. We don't read these plays—they read us. Although Shakespeare was surrounded by literary geniuses, his genius remains unsurpassed.

And yet Jonson was also profoundly wrong about Shakespeare. In many respects, the Bard was entirely of his age, an artist who could only have existed in the London of the late sixteenth century. The point isn't that Shakespeare stole. It's that, for the first time in a long time, there was stuff worth stealing—*and nobody stopped him.* Shakespeare seemed to know this—he was intensely aware that his genius depended on the culture around him. One of the most famous speeches in *Hamlet* comes when the protagonist is addressing a troupe of actors and dispensing advice on their craft. Hamlet urges them to reflect "the very age and body of the time," to transform their surroundings into their "form and pressure." The character is describing the reality of creativity. For Shakespeare, art was inseparable from the whirligig around him, which is why he pilfered his plots and read Italian romances and listened to the feedback of crowds. Shakespeare, of course, was a playwright of unprecedented talent. But that talent was not enough. Shakespeare was for all time, but he could only have existed in his time.[3]

2.

In 1990, the economist Paul Romer invented a new theory of economic growth. Although his theory depends on a long list of abstruse equations, its basic premise is incredibly simple: ideas are an inexhaustible resource. While economics has always been rooted in the scarcity of the material world, Romer pointed out that ideas are a *nonrival good*. When knowledge spreads from person to person, that knowledge isn't diminished or worn out. Instead, ideas tend to become *more* useful when they become more popular—their consumption leads to increasing returns and new innovations. "The fundamental premise of the theory is that there's a big difference between objects and ideas," Romer says. "When we share objects, we make them less valuable. You don't pay as much for a used car because it's already been used. But ideas don't work like that. We can share ideas without devaluing them. There is no inherent scarcity."

Romer's theory comes with very clear implications for creativity. In essence, it suggests that the increased sharing of information is almost always a good thing. "The thing about ideas is that they naturally inspire new ones," Romer says. "This is why places that facilitate idea sharing"—think of, for instance, Silicon Valley or Elizabethan England—"tend to become more productive and innovative than those that don't. Because when ideas are shared, the possibilities do not add up. *They multiply*."

The question, of course, is how to create a multiplier culture.

....

3. *Tragically, the creative flourishing of English literature was brief: in 1642, the Puritans closed down the public theater.*

While cities naturally inspire the sharing of information—all those people can't help but interact—even the most innovative urban areas require help. As Romer points out, Lagos has plenty of human friction, but that friction doesn't lead to new patents—it generates traffic jams that last for days. The favelas in Port-au-Prince are some of the densest settlements in the world, but that density doesn't unleash the potential of its residents—it breeds disease and criminal gangs. "It's important to not be naive about the power of cities," Romer says. "Of course, it's great to concentrate people and give them the freedom to exchange information. But that's not enough. You still need the right set of rules and customs in place to make sure all those people can take advantage of their interactions. You need to have institutions and laws that ensure the costs of density don't outweigh the benefits. Sewage in the streets is never good."

Romer refers to these social concepts as meta-ideas—they are ideas that support other ideas—and he argues that they enable the creativity of a culture. In his academic talks, Romer often shows a photograph of teenagers in Guinea huddled under a streetlamp. At first, it's not clear what the kids are doing or why they're all gathered together in the dark. But then Romer points out the textbooks and pens and calculators—the kids are doing their homework. Although many of these students are wealthy enough to own cell phones, they still live in homes without electricity, which is why they're forced to study in the street. According to Romer, the photograph illustrates the importance of meta-ideas. Because of corruption and price controls in Guinea, most homes remain disconnected from the electrical grid. The country lacks access to a nineteenth-century technology.

But meta-ideas explain more than our societal failings—they also explain our triumphs. Take those ages of excess genius. For too long, we've pretended that these sudden flourishings are mere accidents of history, unworthy of serious investigation. We've told ourselves that they're just periods of freakish genetic gifts, or statistical flukes, which is why we shouldn't bother imitating them. But this is mistake. By looking at the meta-ideas that define a time, we can finally understand why some cultures are so much more creative than others. There is talent everywhere. The only question is whether or not we are taking advantage of it.

Look, for instance, at Elizabethan England. While the period experienced a dramatic increase in knowledge spillovers and urban interaction, this can't fully explain the rise of Shakespeare and his peers. That's because the country didn't simply unleash its bottom-up creativity—it also pioneered a set of meta-ideas that allowed all this creativity to multiply, transforming the density of London into a place of excess genius.

The first important meta-idea developed in sixteenth-century England involved a benign neglect of the rules. During the reign of Queen Elizabeth I, there was an unprecedented relaxation of censorship laws for playwrights. While the government still claimed the right to strictly monitor all speech—in 1581, Elizabeth set up the "master of revels," an official in charge of "regulating" theatrical performances—writers were rarely punished for overstepping the line. This new freedom of expression allowed Shakespeare to criticize the government in his plays, such as his fierce indictment of injustice in *King Lear*:

> *Robes and furred gowns hide all. Plate sin with gold,*
> *And the strong lance of justice hurtless breaks;*
> *Arm it in rags, a pygmy's straw doth pierce it.*

These are the lines of a fearless writer. Shakespeare knew that even if his plays did manage to offend the queen's censors, he probably wouldn't be thrown into a dungeon. (Shakespeare was born into one of the first societies that didn't treat writers like criminals.) Instead, his punishment would be literary; he might be asked to revise the play in the next version, or cut the offending lines from the printed edition. This forgiving attitude encouraged playwrights to take creative risks, to see how much honesty they could get away with. As Shakespeare discovered, the answer was a lot.

Another crucial meta-idea of the time was the concept of intellectual property. While the Elizabethan age had few rules on copyright—Shakespeare was free to steal stories and lines from other writers—it developed new rules for inventions. In the 1560s, the English government began granting exclusive "monopolies for production" to the pioneers of such things as hard soap, glass bottles, and writing paper. Although this patent system was originally designed to lure skilled foreigners, it soon morphed into an important new cultural concept: *ideas have value*. (Francis Bacon declared that the queen would grant a "letter of patent" for any invention deemed useful to the country.) It was no longer enough to protect physical property only—the government also needed to protect intellectual property, so that people had an incentive to invent. For the first time, creativity had become a potential source of wealth.[4]

Perhaps the single most important meta-idea of Elizabethan England involved the spread of education. During the sixteenth century, there was a concerted effort to educate the young males of the middle class, those sons of bricklayers and wool merchants and farmers. A student no longer needed to be vested with a

------

4. *It's at this time that the word* innovation *entered the English language. The noun is derived from the Latin* innovatus, *which roughly translates as "into the new."*

vast estate in order to learn the classics. (Queen Elizabeth's tutor Roger Ascham summarized this modern trend: "All men covet to have their children speak Latin.") Consider the biography of Shakespeare: despite the fact that his father, a glover, was barely literate—John Shakespeare signed his name with a mark—William was sent to the free Stratford grammar school at the age of seven. The school guaranteed an education for most young men, even if their parents were poor and illiterate. (To enforce equality of access, the Oxford-educated schoolmaster was prohibited from accepting payment from his pupils.) While Shakespeare's school day was often tedious, full of rote memorization and corporal punishment, it was also an astonishing opportunity. The son of a rural glover was given access to knowledge that only a few decades before would have been reserved for the privileged few.

And Shakespeare wasn't the only writer to benefit from this new emphasis on public education. Christopher Marlowe was the son of a shoemaker, but he was given a grammar-school education and a full scholarship. The same opportunity was given to Edmund Spenser (son of a London cloth maker) and John Donne (son of an ironmonger). Robert Greene—the same man who would later attack Shakespeare for *not* having a college degree—was the son of poor parents from Norwich, and yet he still managed to obtain a graduate degree from Cambridge. These developments in education led to a vast increase in London's literary talent, expanding the pool of potential playwrights.[5] For Shake-

5. *It's worth pointing out that these "university wits" represented a new demographic in England. Never before had the elite universities been open to such a wide range of students, allowing commoners on scholarships to mingle with the dukes and gentry. Most of these middle-class graduates went on to work in the church. Those who had no interest in God, however, suddenly found themselves with few options and no precursors. (As Stephen Greenblatt notes, "The educational system [of sixteenth-century England] had surged ahead of the existing social system.") The world of commerce was beneath these*

speare, all these new peers and competitors played an essential role in his development. They showed him what was possible. He showed them how it was done.

In a series of essays on Elizabethan literature, T. S. Eliot tried to make sense of the astonishing outpouring of creativity in Elizabethan England. While most critics celebrated the "mythology of Shakespeare"—the man was an outlier who defied all explanation—Eliot focused on the world beyond the writer. He argued that it wasn't an accident that so many famous poets lived in sixteenth-century London, or that the greatest playwright of all time wrote for the same queen as did Marlowe and Jonson. Instead, Eliot believed, these artists were lucky to live in a culture that made it relatively easy to make art. They had been schooled in the traditions of the past but were able to steal from their peers; they knew Latin but wrote in the vernacular; they celebrated complexity in their plays but still managed to sell plenty of tickets. As a result, these Elizabethan writers were able to fulfill their literary potential, transforming a promising time into an age of excess genius. "The great ages did not perhaps produce much more talent than ours," Eliot wrote. "But less talent was wasted."

3.

The New Orleans Center for Creative Arts (NOCCA) is perched on the edge of the Mississippi. Although the public high school is only a few blocks from the souvenir shops of the French Quarter, there are no tourists here, just shotgun homes, hipster

....

educated men—that was the labor of their fathers—but the royal court remained beyond their reach. This left the playhouse, which was uniquely able to straddle these two very different social domains. While the theater companies spent the vast majority of their time performing for the public—the standard ticket cost a penny—many companies also spent several weeks at court every year playing for the queen.

coffee shops, and corner bars. The campus itself is flanked by a freight rail yard and blocks of abandoned factories. This is a neighborhood shadowed by the past tense: that was a praline warehouse; there was the rice mill; this is where they arrested Homer Plessy. If you follow the train tracks to St. Claude and then cross the canal to the east, you end up in the Lower Ninth Ward. The homes there still have high-water marks from the hurricane.[6]

The buses start to arrive at one in the afternoon. The students come to this art school from everywhere; NOCCA draws from the poor parishes of the city and the wealthy suburbs, from the swampy towns of Lake Pontchartrain and the tract homes of Jefferson Parish. Some kids drive all the way from Baton Rouge, ninety miles west. Although the teenagers spend their mornings at "regular school"—their dismissive name for any place that isn't NOCCA—they spend their afternoons here, working on their art.

The students are part of a grand tradition. Since its founding in 1973, NOCCA has graduated an impressive list of artists, including Wynton and Branford Marsalis, Wendell Pierce, Terence Blanchard, Anthony Mackie, Harry Connick Jr., and Trombone Shorty. In 2010, the school sent 98 percent of its seniors to college. While most of these students attended local universities such as Tulane and LSU, NOCCA graduates were also accepted at NYU, UCLA, Juilliard, Northwestern, Oberlin, and the Cooper Union. More than 80 percent of these students received some sort of financial-aid package, with the average student receiving more than $99,000. This means that, although NOCCA costs less than $5

....

6. *After Katrina, the NOCCA campus was used as a staging ground by the Louisiana National Guard for nearly nine months.*

million a year to run, it consistently generates more than $12 million in scholarships per graduating class.

Kyle Wedberg, the CEO of NOCCA, leads me on a tour of the campus. It's a hot spring day, but Wedberg is dressed in a neatly pressed suit. He tells me about the students as we make our way around them, dodging tuba players, a huddle of kids writing sonnets, and an outdoor dance rehearsal. "Many of these kids come from failing schools," Wedberg says. "Many of them have rough family stories or have to work a night job after they're done here. These are not the students who typically get full rides at good colleges. And yet, after a few years at NOCCA, they're ready to go anywhere."

How does NOCCA do it? The process begins with the freshman auditions. Each department holds its own tryouts, requiring applicants to demonstrate their talent. Most of the time, their talent is raw—there are trumpet players who taught themselves how to play and actors who have never taken an acting class. But that doesn't really matter: the teachers are looking for potential, not polish.

Once accepted into NOCCA, the students enter into what seems, at first glance, an antiquated model of education. The school is defined by its master-apprentice approach: the students learn by doing. (Every teacher at NOCCA is also a working artist.) The kids arrive at the school in the early afternoon and then, after scarfing down the free apples and granola bars handed out in the courtyard, head straight to class. What's interesting is what does *not* happen next. The students don't sit in their chairs and listen to a long lecture. (Many rooms at NOCCA don't even have chairs.) They don't retrieve hefty textbooks or begin series of tedious exercises designed to raise their scores on a standardized test. As it turns out, the students don't do any of the things that define the typical high-school experience.

Instead, the students spend their time *creating:* they walk over to their instruments and sketchbooks and costumes and get to work. The campus vibrates with all this imaginative activity; the riffs of jazz echo in the courtyard, and the hallways are flecked with oil paint. The sculpture room smells like wet clay, and the trash cans outside the writing studio are filled to the brim with crumpled sheets of paper. "We're a hundred and twenty years behind the times in all the right ways," Wedberg says. "At some point, vocational education became a dirty word. It became un- fashionable to teach kids by having them do stuff, by having them make things. Instead, schooling became all about giving kids facts and tests. Now, I've got nothing against facts and tests, but memo- rization is not the only kind of thinking we should be encourag- ing." Wedberg pauses, mops the sweat on his brow with his tie, and then leans in as if he is about to confess a secret. "When we obsess over tests, when we teach the way we're teaching now, we send the wrong message to our students," he tells me. "We're basi- cally telling them that creativity is a bad idea. That it's a waste of time. That it's less important than filling in the right bubble. And I can't imagine a worse message than that."

Consider a recent survey of several dozen elementary-school teachers conducted by psychologists at Skidmore College. When asked whether they wanted creative kids in their classroom, every teacher said yes. But when the same teachers were asked to rate their students on a variety of personality measures, the traits most closely aligned with creative thinking (such as being "freely ex- pressive") were also closely associated with their "least favorite" students. Those daydreamers and improvisers might have been imaginative, but they were harder to teach and they underperformed on standardized tests. As a result, they were routinely dismissed and discouraged. The researchers summarized their sad data: "Judgments for the favorite student were negatively correlated with creativity; judgments for the least favorite student were positively correlated with creativity."

The point is that the typical school isn't designed for self-expression; the creative process is often regarded as a classroom failure. "Everyone agrees that creativity is a key skill for the twenty-first century," Wedberg says. "But we're not teaching our kids this skill. We've become so obsessed with rote learning, with making sure that kids memorize the year of some old battle. But in this day and age that's the least valuable kind of learning. That's the stuff you can look up on your phone! If our graduates are going to succeed in the real world, then they have to be able to make stuff. We're a vocational school, but the vocation we care about is creativity."

While NOCCA encourages teenagers to exercise their imagination, recent evidence suggests that it's important to begin this process as early as possible. In one study, researchers compared the mental development of four-year-olds enrolled in a preschool that emphasized unstructured play with those in a more typical preschool in which kids were taught phonetics and counting skills. After a year in the classroom, the students in the play-based school scored better on a variety of crucial cognitive skills, including self-control, the allocation of attention, and working memory. (All of these skills have been consistently linked to academic and real-world achievement.) According to the researchers, the advantage of play is that it's often deeply serious—kids are most focused when they're having fun. In fact, the results from the controlled study were so compelling that the experiment was halted early—it seemed unethical to keep kids in the typical preschool when the play curriculum was so much more effective. As the authors noted, "Unstructured play is often thought frivolous, but it appears to be essential."

Wedberg knows that most NOCCA graduates won't become professional artists. That boy obsessed with the saxophone might become a salesman, and that girl who only wants to tap-dance might become a designer. Nevertheless, these students will still leave the school with an essential talent, which is the ability to develop his or her talent. Because they spend five hours every day working on their own creations, they learn what it takes to get good at something, to struggle and fail and try again. They figure out how to dissect difficult problems and cope with criticism. (One of the defining features of every NOCCA class is the crit session, in which students constructively criticize one another's work.) The students will learn how to manage their own time and persevere in the face of difficulty. "Every kid leaves here with an ability to push themselves," Wedberg says. "We show kids what it takes to make something great."

This is perhaps the most important aspect of NOCCA's project-based curriculum: it exposes students to the brute reality of the creative process. (Remember Milton Glaser's motto: Art Is Work.) It doesn't matter if people are playing jazz or writing poetry—if they want to be successful, they need to learn how to persist and persevere, how to keep on working until the work is done. Woody Allen famously declared that "eighty percent of success is showing up." NOCCA teaches kids how to show up again and again.

In recent years, psychologists have studied the relationship between persistence and creative achievement. They've discovered that the ability to stick with it—the technical name for this trait is *grit*—is one of the most important predictors of success.

"I'd bet that there isn't a single highly successful person who hasn't depended on grit," says Angela Duckworth, a psychologist at the University of Pennsylvania who helped pioneer the study of the psychological trait. "Nobody is talented enough to not have to work hard, and that's what grit allows you to do." Duckworth has found that levels of grit predict success at the National Spelling Bee, graduation from the Special Forces boot camp, and even teacher effectiveness in the Teach for America program. (In many instances, grit explains a greater percentage of individual variation than do intelligence and IQ scores.) "What grit allows you to do is to take advantage of your potential," Duckworth says. "Because even the smartest, most talented people still need to practice. If you're a violin player, grit is what gets you to keep on practicing, even when the practice isn't very fun. If you're a novelist, grit allows you to finally finish that first novel. And then it lets you keep on working on the novel until it's actually good."

The vocational approach at NOCCA helps build grit in students. It teaches them how to be single-minded in pursuit of a goal, to sacrifice for the sake of a passion. The teachers demand hard work from their kids because they know, from personal experience, that creative success requires nothing less. While most people assume that art schools are somehow lacking in rigor—"They think we just let the students screw around," says one NOCCA administrator—the irony is that these schools do a much better job of endowing their apprentices with the most essential mental skills. In one painting class, I watched as a teacher dissected the work of a level 1 student. (At NOCCA, kids are grouped by skill level, not age or grade.) "This is an interesting idea," she said while reviewing the brightly colored still life. "But you didn't execute. I can tell that you rushed a little bit—your brushstrokes are all over the place. This could have been very good. But you needed to put in more work." And so the fifteen-year-old is given yet another lesson in grit. He is reminded, once again, that creativity is damn hard.

NOCCA isn't for everyone, of course. Some students aren't interested in making art or performing onstage. Some teenagers don't want to stay at school until dark or commute for hours on the bus. Nevertheless, the guiding principles of NOCCA—that creativity can be taught, and that our kids are reservoirs of untapped talent—deserve to be widely implemented. Although school reform typically focuses on the bottom quartile of students, we shouldn't let this concern for those at risk of dropping out blind us to the importance of encouraging excellence. "Our goal here is to create a pocket of brilliance," Wedberg says. "Most of our students would do fine if they were stuck in a regular school. They'd get decent grades. They'd probably even go to some kind of college. But we shouldn't be satisfied with that. We should insist that they live up to their potential. Because it's not enough to be good when you can be great."

And it's not just art schools that are creating pockets of brilliance. High Tech High, a San Diego charter school, was founded in 2000 by a group of local tech executives frustrated by a lack of skilled workers. The school renovated an abandoned navy facility, transforming the cavernous concrete warehouses into a series of loft-like classrooms. Like NOCCA, High Tech High emphasizes learning by doing—every student is required to complete numerous projects that take up much of his or her school day. (Instead of playing with paint and guitars, the kids play with scrap metal and programming code.) "People act like we've got this radical concept of education," says Larry Rosenstock, the fast-talking CEO of the school. "But it's actually been around for a long time. [John] Dewey said it best: 'Understanding derives from activity.' Kids don't learn when they're consuming information, when someone is talking down to them. They learn when they're producing stuff. That's how you get them to work hard without realizing they're working." Past projects at High Tech High have included the construction of a human-powered submarine and a robot capable of accurately kicking a soccer ball forty-five feet. Last year, a High Tech High art class spent several weeks retrofitting an old cigarette vending machine — it cost two hundred dollars on eBay — so that it could sell student paintings. The machine is now set up in the lobby.

While High Tech High was initially criticized for its emphasis on these projects — some argued that students needed more classroom instruction — their academic results are indisputable. Every single High Tech High graduate has been admitted to college, and more than 85 percent of these students have graduated from four-year institutions. Furthermore, nearly a third of these graduates are first-generation college students. (High Tech High admits students using a random lottery system, drawing freshmen from all over the San Diego area.) "Our kids are growing up in a world of constant change," Rosenstock says. "There is no test for the future that we can teach to. What we do know, however, is that being able to make new things is still going to be the way to succeed. Creativity is a skill that never goes out of style."

What schools like NOCCA and High Tech High demonstrate is that the imagination is too important to be ignored. When children are allowed to create, they're able to develop the sophisticated talents that are required for success in the real world. Instead of learning how to pass a standardized test, they learn how to cope with complexity and connect ideas, how to bridge disciplines and improve their first drafts. These mental talents can't be taught in an afternoon — there is no textbook for ingenuity, no lesson plan for divergent thinking. Rather, they must be discovered: the child has to learn by doing.

This was demonstrated in a recent study led by MIT psychologist Laura Schulz. The experiment consisted of giving four-year-olds a new toy outfitted with four tubes. What made the toy interesting is that each tube did something different. One tube, for instance, generated a squeaking sound, while another tube turned into a tiny mirror.

The first group of students were shown the toy by a scientist who declared that she'd just found it on the floor. As she revealed the toy to the kids, she "accidentally" pulled one of the tubes and made it squeak. Her response was sheer surprise: "Huh! Did you see that? Let me try to do that again!" The second group got a very different presentation. Instead of feigning surprise, the scientist acted like a typical teacher. She told the students that she'd gotten a new toy and that she wanted to show them how it worked. Then she deliberately made the toy squeak.

After the demonstration, both groups of children were given the toy to play with. Not surprisingly, all of the children pulled on the first tube and laughed at the squeak. But then something interesting happened. While the children from the second group quickly got bored with the toy, those in the first group kept on playing with it. Instead of being satisfied with the squeaks, they explored the other tubes and discovered all sorts of hidden surprises. According to the psychologists, the different reactions were caused by the act of teaching. When students are given explicit instructions, when they are told what they need to know, they become less likely to explore on their own. Curiosity is a fragile thing.

That's why the best schools ensure that unstructured play — what happens when the child creates and explores on her own — is an essential part of the classroom experience. It doesn't matter if the student is writing a poem or soldering a computer circuit or scribbling with crayons — she needs to feel for herself the thrill and struggle of making something new. Because even if these students at NOCCA and High Tech High don't end up pursuing their art or majoring in computer science, they will never forget what they learned as teenage artists and engineers.

On my last day at NOCCA, I spent a few hours in the auditorium surrounded by five hundred extremely excited students. The kids were getting ready for interlude day, their chance to perform for one another. The range of expression at the Interlude was stunning. Although the show began with someone singing Brahms, it quickly veered into poetry and jazz guitar. There were PowerPoint slides of oil paintings and a collection of funny video shorts; someone recited a *Macbeth* monologue and a troupe of students performed a scene from *West Side Story*. The only connecting thread was the ardor of the kids onstage. After the performances were over — the students gave themselves a standing ovation — I struck up a conversation with Tiffani, a dance student. I asked her if she planned on becoming a professional dancer. "Probably not," she replies. "I love to dance — it makes me so happy — but dancers make no money. I want to make money." I then ask Tiffani if she thinks her dance training will still be useful. Wouldn't it be better to go to a regular school? "Oh, no way," she says. "I'm not just learning how to dance here. It might look like that when you look at our classes because we're always dancing. But that's not it. What I'm really learning is how to say something."

4.

What are the meta-ideas that *we* need to embrace? How can we create more pockets of brilliance? This might seem like an impossible task, a misguided attempt to replicate a vanished golden age. But it's not. We now have enough evidence to begin prescribing a set of policies that can increase our collective creativity. In fact, we've already proven that it's possible to create a period of excessive genius, a moment that's overflowing with talent. The only problem is that the geniuses we've created are athletes.

Bill James, the pioneer of sabermetrics—the statistical analysis of baseball—points out that modern America is prodigiously good at producing sports stars. As a result, a city like Wichita, Kansas—roughly the same size as Elizabethan London—can produce a professional athlete every few years. Think about how impressive that is: the high schools of Wichita are able to regularly churn out talented individuals such as Barry Sanders and Gayle Sayers, capable of competing at the highest levels in the world. Their physical genius—which is often quite creative—is worth millions of dollars.

And yet the same excess does not apply to other kinds of talent. Wichita has not produced a surplus of gifted writers, painters, jazz musicians, or inventors. As James notes, this is largely because our culture treats athletes differently. The first thing we do is encourage them when they're young, driving the kids to baseball practice and Pop Warner tournaments. This doesn't just allow children to develop their talent—it also lets coaches identify those with the most natural ability. Second, we constantly celebrate athletic success. Winning teams get trophies and parades, coverage in the local newspaper, and the congratulations of the community. Finally, we have mechanisms for cultivating those with athletic potential at every step of the process, from Little League to the NCAA to the major leagues. They are showered with attention and rewarded with huge contracts.

So it's possible to create more geniuses—we've already done

it. The question now is whether our society can produce creative talent with the same efficiency that it has produced athletic talent. Our future depends upon it.

The first meta-idea we need to take seriously is education. Because it's impossible to predict where the next genius will come from, ages of excessive genius are always accompanied by new forms of educational opportunity. They occur when the sons of illiterate tradesmen go to college, when even the child of a glover gets a library. Like in Elizabethan England, we need to ensure that every student has a chance to succeed. We do an excellent job of lavishing gifted athletes with attention and scholarships, but too many of their peers are forced to attend failing schools with high dropout rates. Their imaginations never have a chance. Think of all the wasted potential.

However, it's not enough to increase access to the classroom or raise the test scores of the lowest performers. We also have to ensure that those with talent are allowed to flourish, that we have institutions that can nourish our brightest kids, just as we nourish our best quarterbacks and jump shooters. Like the administrators at NOCCA and High Tech High, we must identify those with motivation and potential and then give them the tools to discover and invent. "Something very special happens when you concentrate talent," Wedberg says. "The students here inspire and challenge each other. My favorite moments are when I see kids who are surprised by what they've done. It's like they can't believe they're actually this good."

If we're not going to properly educate our own children, then we need to at least open the doors and encourage immigration. This is the second important meta-idea: ages of excess genius are always accompanied by new forms of human mixing. The numbers are persuasive. According to the latest figures from the U.S.

Patent Office, immigrants invent patents at double the rate of non-immigrants, which is why a 1 percent increase in immigrants with college degrees leads to a 15 percent rise in patent production. (In recent years, immigrant inventors have contributed to more than a quarter of all U.S. global patent applications.) These new citizens also start companies at an accelerated pace, cofounding 52 percent of Silicon Valley firms since 1995.[7] We all benefit when those with good ideas are allowed to freely move about.

Just look at Elizabethan England, which experienced an unprecedented mixing of its population. Some of this mixing was born of urban density, as people flocked to London from all over the country. However, the period was also marked by the rise of international trade and the emergence of a merchant class that moved freely across national borders. "What you see in this period is a dramatic growth in the number and variety of human collisions," says Robert Watson, a professor of English literature and history at UCLA. "People were meeting people like never before."

One of the consequences of all these new "collisions" was an explosion of new words in the English language. Some of these words came from the city streets, as all the recently arrived Londoners were forced to reconcile their regional dialects and local idioms. However, most of the novel words came from abroad. "What really seems to be driving the growth [of the language] is this large group of multilingual citizens in England," says Watson. "You've got people learning French, Latin, Greek, and German.

---

7. *Last, immigrants bring America a much-needed set of skills and interests. In 2010, foreign students studying on temporary visas received more than 60 percent of all U.S. engineering doctorates. (American students, by contrast, dominate doctorate programs in the humanities and social sciences.) What makes these engineering degrees so valuable is that, according to the Department of Labor, the 5 percent of American workers employed in fields related to science and technology are responsible for more than 50 percent of sustained economic growth.*

They're traveling abroad, encountering new things. And they can't help but import many of these foreign sayings into English." As a result, the writers of the period had a vastly expanded palette of expressions with which to paint their world. Shakespeare, for one, took advantage of it: his work features a vocabulary that's unparalleled in literature, as his plays use more than twenty-five thousand different words. (His closest rival in terms of variety was John Milton, who clocked in with less than half that.) The richness of Shakespeare's art is inseparable from this richness of language, which itself depended on those immigrants around him.

This lesson isn't restricted to the sixteenth century—encouraging the collisions of creative people is always a good idea. In fact, these interactions are so important that even seemingly minor regulations can have an outsize effect. Consider the presence of noncompete clauses, those binding contracts that prevent employees from working for competitors. Thanks to a quirk of the California Civil Code, virtually all noncompete clauses are void in the state. As a result, engineers in Silicon Valley are free to constantly jump between firms and chase more interesting problems and bigger paychecks. This leads, over the long run, to a surplus of horizontal interactions and weak ties. (According to a recent analysis by the Federal Reserve, the unbridled movement of workers in California has played an important role in the development of Silicon Valley.) The larger point is that we meddle with the social network at our own peril. Like Shakespeare, we should aspire to live in a time filled with new words.

Another crucial meta-idea is a willingness to take risks. It doesn't matter if we're giving out small-business loans or research grants to young scientists: we have to consistently encourage those who take chances. Most entrepreneurs will fail, and many of those grants will lead to inconclusive experiments. (Even Shakespeare wrote a number of bad plays.) But those failures are a sign that the system is working, that we're giving new ideas a chance. In my conversations with Yossi Vardi, the start-up impresario of Tel Aviv, he repeatedly referred to the importance of *chutzpah,* the Yiddish word for audacity. "You can't have creativity without chutzpah," he says. "It takes enormous chutzpah to believe that you have an idea that will change the world and make a lot of money. But unless you believe that, then you will never become an entrepreneur."

The reason chutzpah is so important has to do with the nature of new ideas, which are inherently precarious. As AnnaLee Saxenian notes, most successful entrepreneurs in Silicon Valley have failed with at least one previous start-up. Their failure, however, doesn't prevent them from trying again. And again. Or look at NOCCA: the students at the school are always encouraged to take risks, to experiment with the possibility of embarrassment. When I walked into the classroom of Silas Cooper, a drama teacher at the school, I couldn't help but notice the handwritten banner hanging above the door. This is what it said: FAIL BIG.

One way to illustrate the importance of encouraging risk is to compare the research strategies of the National Institutes of Health (NIH)—the largest funder of biomedical science in the world—and the Howard Hughes Medical Institute (HHMI), a large nonprofit set up to "push the boundaries of knowledge." The NIH evaluates grant proposals in an exceedingly rational manner. A team of experts analyzes and scores each proposal to ensure that the project is scientifically sound and supported by plenty of preliminary evidence. Their explicit goal is to not waste taxpayer money—nobody wants to fund a failure.

HHMI, in contrast, is known for supporting avant-garde projects. In fact, it explicitly encourages researchers to "take risks, explore unproven avenues and embrace the unknown—even if it means uncertainty or the chance of failure." HHMI does this by focusing on individual scientists, not particular experiments. (Instead of requesting a detailed proposal of future research, HHMI asks for an example of past research.) The assumption is that a creative scientist should be able to pursue ideas without having to justify them to a panel of experts. Sometimes the experiments with the most potential are still lacking evidence.

A few years ago, a team of economists at MIT and UCSD analyzed the data from NIH- and HHMI-funded labs to see which funding strategy was more effective. The economists tried to control for every possible variable, such as outside scholarships and the quality of graduate students. Then they compared the output of NIH researchers to HHMI investigators with similar track records.

The data was clear: in every biomedical field, the risky HHMI grants were generating the most important, innovative, and influential research. Although HHMI researchers had qualifications similar to their NIH counterparts when they first applied for funding, they went on to produce twice as many highly cited research articles and win six times as many awards. They also introduced more new keywords into the scientific lexicon, which is a marker of highly original work.

The bad news, of course, is that all this creativity comes with a cost. This is why, according to the economists, the HHMI researchers also produced 35 percent more research papers that were cited by nobody at all. (These papers were abject failures.) The moral is that these scientists weren't producing better research because they were smarter or more creative or had more money. Instead, they had more success because they were more willing to fail.

Bill James makes a similar point in terms of sports. He notes that American society has found a way to value athletic potential and not just achievement: "We invest in athletes that might be good, rather than simply paying them once they get to be among the best in the world." Of course, not all of these prospects work out. Some draft picks are busts, and many highly paid players disappoint. (Athletes, in other words, are a lot like scientific grants.) Nevertheless, professional teams realize that this system is necessary, since it encourages young athletes to pursue the sport, to invest the time and energy needed to succeed. Betting on potential is always a risk, but that's the only way to get a surplus of talent. And that's why we need more foundations and government-funding agencies willing to imitate the bold model of Howard Hughes and professional sports teams.

The final essential meta-idea involves managing the rewards of innovation. Inventors should profit from their past inventions, but we also need to encourage a culture of borrowing and adaptation. This tension has been present ever since Queen Elizabeth began granting patents in the late sixteenth century. Although patents serve as an important source of motivation—Abraham Lincoln described the patent system as "adding the fuel of interest to the fire of genius"—they also make it harder for other inventors to build on the innovation. This is why, in 1601, the English government began revoking many of the patents most despised by the public, including those on glass bottles and starch. What the queen discovered is that there is nothing natural about the scarcity of ideas. Of course, just because ideas want to be free doesn't mean they *should* be free. It just means that we have to get the price right.

Unfortunately, that isn't happening. In recent years, American creativity has been undermined by an abundance of vague patents and the recurring extension of copyright claims. Let's begin with patents. Between 2004 and 2009, patent-infringement lawsuits increased by 70 percent, while licensing-fee requests rose by 650 percent. Many of these lawsuits were brought by so-called patent trolls, those individuals and firms who buy patents in bulk and then aggressively hunt for possible infringements even though they have no interest in using the patented inventions. Or consider the length of copyright protection: when the first copyright laws were passed in 1790, the length of protection was fourteen years. (As Lewis Hyde notes, the Founding Fathers were deeply invested in the notion that practically all created works should belong to "the commons."[8]) Since 1962, however, Congress has extended copyright protection eleven times, and the typical length of protection is now ninety-five years. The problem with these extensions is that they discourage innovation, preventing people from remixing and remaking old forms. There will always be a powerful business lobby for the protection of intellectual property—the 1998 copyright extension law was nicknamed the Mickey Mouse Protection Act—but we need to remember that the public domain has no lobby. And that's why we should always think of young William Shakespeare stealing from Marlowe and Holinshed and Kyd. (If Shakespeare were writing today, his plays would be the subject of endless lawsuits.) It doesn't matter if it's a hip-hop album made

....

8. *One of America's initial economic advantages was the weak grip of trade guilds. The primary function of these guilds, which dominated fields ranging from printing to soap making, was the restriction of technical knowledge. However, the relative weakness of guilds in the American Colonies meant that much of this valuable knowledge was widely disseminated, cheaply available to the public in the form of reference manuals and trade books. For instance, when Ben Franklin became publisher of the* Philadelphia Gazette, *in 1729, he declared that one of the main goals of the paper would be the dissemination of "such Hints . . . as may contribute either to the Improvement of our present Manufactures, or towards the Invention of new Ones."*

up of remixes and music samples or an engineer tweaking a gad-get in a San Jose garage: we have to make sure that people can be inspired by the work of others, that the commons remains a rich source of creativity.

Bob Dylan illustrates this point beautifully. In *Chronicles,* his autobiography, Dylan repeatedly describes his creative process as one of love and theft. The process begins when he finds a sound or song that "touches the bone." He then tries to deconstruct the sound to figure out how it works. When Dylan was a young song-writer in New York City, for instance, he learned to write music by memorizing his influences, studying the melodic details of Rob-ert Johnson, Woody Guthrie, and a long list of English folk bal-lads. (Dylan, in *Chronicles:* "I could rattle off all these songs with-out comment as if all the wise and poetic words were mine and mine alone.") But Dylan wasn't just copying these tunes; his close study was an essential part of his creative method—learning an old song meant that he was on the verge of inventing a new one. Dylan describes how this worked in one of his first recording ses-sions:

> I didn't have many songs, but I was rearranging verses to old blues ballads, adding an original line here or there, anything that came into my mind—slapping a title on it.

One consequence is that virtually all of Dylan's first seventy compositions, from "Blowin' in the Wind" to "The Times They Are a-Changin'," have clear musical precursors. In most instances, the original folk composition is obviously there, a barely con-cealed inspiration. While it would be easy to dismiss such songs as mere rip-offs—several of them would almost certainly violate current copyright standards—Dylan was able to transform his folk sources into pop masterpieces. T. S. Eliot said it best: "Immature poets imitate. Mature poets steal." Even at the age of twenty-one, Dylan was a mature poet. He was already a thief.

These are the meta-ideas that have worked before, unleashing the talent of past playwrights and inventors. They are the top-down policies that have set free our bottom-up creativity. But this list is not complete, not even close. As Romer notes, "We do not know what the next major idea about how to support ideas will be." And this is why it's so important to keep searching for the effective meta-ideas of the future, for the next institution or attitude or law that will help us become more creative. We need to innovate innovation.

Because here is the disquieting truth: Our creative problems keep on getting more difficult. Unless we choose the right policies and reforms, unless we create more NOCCAs and fix the patent system, unless we invest in urban density, unless we encourage young inventors with the same fervor that we encourage young football stars, we'll never be able to find the solutions that we so desperately need. It's time to create the kind of culture that won't hold us back.

The virtue of studying ages of excess genius is that they give us a way to measure ourselves. We can learn from the creative secrets of the past, from those outlier societies that produced Shakespeare and Plato and Michelangelo. And then we should look in the mirror. What kind of culture have we created? Is it a world full of ideas that can be connected? Are we willing to invest in risk takers? Do our schools produce students ready to create? Can the son of a glover grow up to write plays for the queen? We have to make it easy to become a genius.

CODA

In the summer of 1981, Penn and Teller were struggling magicians plying their trade on the Renaissance Faire circuit, performing a set of middling tricks for little kids. Their costumes were embarrassing, getups of black tights, purple velour capes, fake leather vests, and belts made of rope. They were frustrated and fed up, tired of life on the road. "I was definitely on the verge of giving up the dream of becoming a magician," Teller says. "I was ready to go back home and become a high-school Latin teacher."

But then, just as despair began to overtake the magicians, a breakthrough arrived in a highway diner. While waiting for his food to arrive, Teller wanted to practice his version of cups and balls, a classic sleight-of-hand trick first performed by conjurers in ancient Rome. The magician begins by placing three small balls underneath three cups. Then the magician engages in a series of vanishes and transpositions, making the balls repeatedly appear and disappear. Just when the spectator assumes he knows where the ball is — it's inside the cup! — the ball is revealed to be elsewhere. "This trick is performed all over the world," Penn says. "If you see a guy hustling on the street, he's probably doing cups and balls."

But Teller was in a diner—he wasn't carrying his bag of magic supplies. And so he used what he had: rolled-up napkins for the balls, and clear plastic water glasses as cups. The epiphany arrived halfway through the trick. Although it was now possible to follow the crumpled napkins from cup to cup, Teller realized that the illusion persisted. "The eye could see the moves, but the mind could not comprehend them," he says. "Giving the trick away gave nothing away, since people still couldn't really grasp it." Because watchers were literally incapable of perceiving the sleight-of-hand—Teller's fingers just moved too fast—it didn't matter if the glasses were see-through and the napkins were visible.

Penn and Teller worked this version of cups and balls into their traveling show. It was a huge hit. Before long, they were performing with clear plastic glasses on *Letterman* and playing to sold-out crowds in New York City. In the decades since, Penn and Teller have become two of the most successful magicians in the world—they now have their own theater in Vegas and a secret warehouse off the Strip where they develop new tricks. ("I no longer have to find my ideas in diners," Teller says.) Nevertheless, this popularity hasn't diminished their avant-garde spirit: Penn and Teller are still determined to deconstruct their own magic. In their current show, for instance, Penn frequently shouts out the secret while Teller is performing the illusion: "This is just invisible string!" he might say, or "It's only a mirror!" According to the magicians, their "skeptical shtick" can be traced back to that diner when Teller was forced to make magic with the only things around. "In many respects, that simple idea of clear glasses has come to represent what we're trying to do," Teller says. "The reason I will always love our version [of cups and balls] is that, even when you give away the trick—you hide nothing—the magic is still there. In fact, the illusion becomes even more meaningful, because you realize that it's all in your head. There is nothing special about these glasses and napkins. The magic is coming from your mind."

Creativity is like that magic trick. For the first time, we can see the source of imagination, that massive network of electrical cells that lets us constantly form new connections between old ideas. However, this new knowledge only makes the act itself more astonishing. The moment of insight might emanate from an obscure circuit in the right hemisphere, but that doesn't diminish the thrill of having a new idea in the shower. Just because we can track the flux of neurotransmitters, or measure the correlation between walking speed and the production of patents, or quantify the effect of the social network, that doesn't take away from the sheer wonder of the process. There will always be something slightly miraculous about the imagination.

Nevertheless, this sense of magic shouldn't prevent us from trying to become more creative. Thanks to modern science, we've been blessed with an unprecedented creative advantage, a meta-idea that we can all apply at the individual level. For the first time in human history, it's possible to learn how the imagination actually works. Instead of relying on myth and superstition, we can think about dopamine and dissent, the right hemisphere and social networks. This self-knowledge is extremely useful knowledge; because we've begun to identify the catalysts behind our creativity, we can make sure that we're thinking in the right way at the right time, that we're fully utilizing this astonishing tool inside the head. (There is no more important meta-idea than knowing where every idea comes from.) If we want to increase our creative powers, then we have to put this research to work in our own lives. We can imagine more than we know.

The process begins with the brain, that fleshy source of possibility. Although people have long assumed that the imagination is a single thing, it's actually a talent that takes multiple forms. Sometimes we need to relax in the shower and sometimes we need to chug caffeine. Sometimes we need to let ourselves go, and sometimes we need to escape from what we know. There is a time for every kind of thinking.

But the brain is only the beginning. We now know creativity is also an emergent property of people coming together. When collaborating with others, we should seek out the sweet spot of Q, just like the artists behind *West Side Story*. Brainstorming might feel nice, but constructive criticism is always better; every company needs the equivalent of the Pixar bathroom, a space that forces employees to interact as if they were in a dense city. And then we need to get our meta-ideas right so that we don't inhibit our collective imagination. We should aspire to excessive genius.

But we must also be honest: the creative process will never be easy, no matter how much we know about neurons and cities and Shakespeare. Our inventions will always be shadowed by uncertainty and contingency, by the sheer serendipity of brain cells making new connections. The science of the imagination doesn't fit neatly on a PowerPoint slide, and it can't be summarized in a subtitle. (If creativity were that easy, Picasso wouldn't be so famous.) Despite all the clever studies and rigorous experiments, our most essential mental talent remains the most mysterious.

The mystery is this: although the imagination is inspired by the everyday world—by its flaws and beauties—we are able to see beyond our sources, to imagine things that exist only in the mind. We notice an incompleteness and we can complete it; the cracks in things become a source of light. And so the mop gets turned into the Swiffer, and Tin Pan Alley gives rise to Bob Dylan, and a hackneyed tragedy becomes *Hamlet*. Every creative story is different. And every creative story is the same. There was nothing. Now there is something. It's almost like magic.

NOTES

Introduction

xii   *"I think P and G"*: Swiffer team, interview with the author, Continuum headquarters, Newton, Massachusetts, April 5, 2009; Harry West, follow-up phone interview with author, January 21, 2010.

1. Bob Dylan's Brain

3   *Bob Dylan looks*: For a behind-the-scenes look at Dylan's 1965 British tour, D. A. Pennebaker's documentary *Don't Look Back* remains without parallel. Other valuable secondary sources on the writing of "Like a Rolling Stone" include Greil Marcus's *Like a Rolling Stone: Bob Dylan at the Crossroads* (New York: Public Affairs, 2005); *The Bob Dylan Encyclopedia*, by Michael Gray (New York: Continuum, 2008); and *No Direction Home*, by Robert Shelton (New York: Ballantine Books, 1987).

5   *"I realized I"*: Nat Hentoff, "Interview with Bob Dylan," *Playboy*, February 1966.

8   *"The doctors would"*: Mark Beeman, personal interviews with the author, Northwestern University, April 10 to 12, 2008.

    the right hemisphere: http://nobelprize.org/nobel_prizes/medicine/laureates/1981/sperry-lecture.html.

11   *Schooler presented*: Jonathan Schooler, Stellan Ohlsson, and Kevin

Brooks, "Thoughts Beyond Words: When Language Overshadows Insight," *Journal of Experimental Psychology* 122 (1993): 166–83.

12 *"One of the"*: Jonathan Schooler, telephone interview with the author, April 6, 2008.

16 *By combining both*: John Kounios et al., "The Prepared Mind: Neural Activity Prior to Problem Presentation Predicts Solution by Sudden Insight," *Psychological Science* 17 (2006): 882–980; Mark Jung-Beeman et al., "Neural Activity Observed in People Solving Verbal Problems with Insight," *Public Library of Science — Biology* 2 (2004): 500–10.

19 *"I found myself"*: Bob Dylan, radio interview with Martin Bronstein, Canadian Broadcasting Company, February 20, 1966.

24 *"Last spring, I guess"*: Hentoff, "Interview."

2. Alpha Waves (Condition Blue)

28 *"We're an unusual"*: Larry Wendling, personal interview with the author, 3M headquarters, February 2, 2009.

30 *Joydeep Bhattacharya, a psychologist*: Simone Sandkühler and Joydeep Bhattacharya, "Deconstructing Insight: EEG Correlates of Insightful Problem Solving," *PLoS ONE* 3; http://www.plosone.org/article/info:doi/10.1371/journal.pone.0001459.

32 *German researchers have found*: Annette Bolte, Thomas Goschke, and Julius Kuhl, "Emotion and Intuition: Effects of Positive and Negative Mood on Implicit Judgments of Semantic Coherence," *Psychological Science* 14 (2003): 416–22.

34 *Consider an experiment*: Carlo Reverberi et al., "Better Without (Lateral) Frontal Cortex? Insight Problems Solved by Frontal Patients," *Brain* 128 (2005): 2882–90.

35 *Their attention deficit*: Holly White and Priti Shah, "Creative Style and Achievement in Adults with Attention-Deficit/Hyperactivity Disorder," *Personality and Individual Differences* 50 (2011): 673–77.

38 *Mark Turner, a cognitive*: Mark Turner, editor, *The Artful Mind: Cognitive Science and the Riddle of Human Creativity* (Oxford: Oxford University Press, 2006), 107–10.

39 *"All this creative"*: David Hume, *An Enquiry Concerning Human Understanding* (1777; reprint Oxford: Oxford University Press, 2001), 11.

43 *According to Mary Gick*: Mary Gick and Keith Holyoak, "Analogical Problem Solving," *Cognitive Psychology* 12 (1980): 306–55.

The importance of: Shelley Carson, Jordan Peterson, and Daniel Higgins, "Decreased Latent Inhibition Is Associated with Increased Creative

Achievement in High-Functioning Individuals," *Journal of Personality and Social Psychology* 85 (2003): 499–506.

44 *"We told the"*: Marcus Raichle, telephone interview with the author, May 18, 2009.

46 *"Certainly she was"*: Virginia Woolf, *To the Lighthouse* (1927; reprint Boston: Mariner Books, 2005), 163.

48 *"What these tests"*: Jonathan Smallwood, Jonathan Schooler, and Todd Handy, "Going AWOL in the Brain: Mind Wandering Reduces Cortical Analysis of External Events," *Journal of Cognitive Neuroscience* 20 (2008): 458–69.

49 *This helps explain*: Michael Sayette, Erik Reichle, and Jonathan Schooler, "Lost in the Sauce: The Effects of Alcohol on Mind Wandering?," *Psychological Science* 20 (2009): 747–52.

51 *Look at this*: Ravi Mehta and Rui Zhu, "Blue or Red? Exploring the Effect of Color on Cognitive Task Performances," *Science* 323 (2009): 1226–29.

3. The Unconcealing

53 *And so, before*: Stan Smith, editor, *The Cambridge Companion to W. H. Auden* (Cambridge: Cambridge University Press, 2005), 22–25; Richard Davenport-Hines, *Auden* (New York: Pantheon, 1996), 185–87.

65 *"There were machines"*: Earl Miller, interview with the author, MIT, April 2, 2008.

68 *"I was just"*: Milton Glaser, telephone interviews with the author, March 20, 2009, and March 21, 2009.

73 *The only way*: For a lucid summary of Heidegger's thinking on unconcealment, see Mark Wrathall, *Heidegger and Unconcealment: Truth, Language, and History* (Cambridge: Cambridge University Press, 2010).

76 *Joe Forgas, a social*: Joseph Forgas, "When Sad Is Better Than Happy: Negative Affect Can Improve the Quality and Effectiveness of Persuasive Messages and Social Influence Strategies," *Journal of Experimental Social Psychology* 43 (2007): 513–28.

77 *In one of*: Modupe Akinola and Wendy Mendes, "The Dark Side of Creativity: Biological Vulnerability and Negative Emotions Lead to Greater Artistic Creativity," *Personality and Social Psychology Bulletin* 34 (2008): 1677–86.

78 *According to her data*: Nancy Andreasen, "Creativity and Mental Illness: Prevalence Rates in Writers and Their First-Degree Relatives," *American Journal of Psychiatry* 144 (1987): 1288–92.

79 *"Successful writers are"*: Nancy Andreasen, telephone interview with author, December 10, 2009.

80 *More recently, the*: Hagop Akiskal and Kareen Akiskal, "In Search of Aristotle: Temperament, Human Nature, Melancholia, Creativity and Eminence," *Journal of Affective Disorders* 100 (2007): 1–6.

82 *This was first*: Janet Metcalfe, "Feeling of Knowing in Memory and Problem Solving," *Journal of Experimental Psychology* 12 (1986): 288–94; Janet Metcalfe and David Wiebe, "Intuition in Insight and Noninsight Problem Solving," *Memory and Cognition* 15 (1987): 238–46.

4. The Letting Go

84 *"There's always that"*: Bruce Adolphe, interview with the author, American Museum of Natural History, New York City, May 3, 2009.

86 *"I was nineteen"*: David Blum, "A Process Larger Than Oneself," *The New Yorker*, May 1, 1989.

87 *"If you are"*: Yo-Yo Ma, telephone interview with the author, January 10, 2011.

89 *Charles Limb, a neuroscientist*: Charles Limb and Allen Braun, "Neural Substrates of Spontaneous Musical Performance," *PLoS ONE* 3 (2008); e1679.doi:10.1371/journal.pone.0001679.

92 *That was the*: Aaron Berkowitz and Daniel Ansari, "Generation of Novel Motor Sequences: The Neural Correlates of Musical Improvisation," *Neuroimage* 41 (2008): 535–43.

95 *"You have no idea"*: Mitch Varnes, interview with the author, Maui, May 16, 2009.

96 *In one study*: J. Craig and Simon Baron-Cohen, "Creativity and Imagination in Autism and Asperger Syndrome," *Journal of Autism and Developmental Disorders* 29 (1999): 319–26.

97 *"I don't know"*: Clay Marzo, interviews with the author, Maui, May 16–19, 2009.

101 *"The hardest part"*: Joshua Funk, interviews with the author, Hollywood, March 3–4, 2010.

106 *Bruce Miller is*: Bruce L. Miller, "Emergence of Artistic Talent in Frontotemporal Dementia," *Neurology* 51 (1998): 978–82.

107 *Take a 2004 paper*: Ullrich Wagner and Jan Born, "Sleep Inspires Insight," *Nature* 427 (2004): 352–55.
Or consider a: Denise Cai et al., "REM, Not Incubation, Improves Creativity by Priming Associative Networks," *PNAS* 106 (2009): 10130–34.

109 *Allan Snyder, a neuroscientist*: Allan Snyder, "Savant-Like Skills Exposed

in Normal People by Suppressing the Left Fronto-Temporal Lobe," *Journal of Integrative Neuroscience* 2 (2003): 149–58.

109 *According to Snyder:* Allan Snyder, "Explaining and Inducing Savant Skills: Privileged Access to Lower Level, Less-Processed Information," *Philosophical Transactions of the Royal Society* 364 (2009): 1399–1405.

110 *It's at this:* E. Paul Torrance, "A Longitudinal Examination of the Fourth Grade Slump in Creativity," *Gifted Child Quarterly* 12 (1968): 195–99.
Take this clever: Darya Zabelina and Michael Robinson, "Child's Play: Facilitating the Originality of Creative Output by a Priming Manipulation," *Psychology of Aesthetics, Creativity, and the Arts* 4 (2010): 57–65.

5. The Outsider

112 *"It was a":* Don Lee, interview with the author, Culver City, November 28, 2009, and December 14, 2009.

118 *"After spending years":* Alpheus Bingham, telephone interview with the author, December 12, 2009.

120 *According to Lakhani's:* Karim Lakhani, "The Value of Openness in Scientific Problem Solving," *Technology and Operations Management;* http://www.hbs.edu/research/pdf/07-050.pdf.

122 *"It was really":* Jeff Howe, "The Rise of Crowdsourcing," *Wired* (June 2006): 81–89.

123 *Dean Simonton, a psychologist:* Dean Keith Simonton, *Origins of Genius: Darwinian Perspectives on Creativity* (Oxford: Oxford University Press, 1999).

124 *"start to repeat":* Dean Simonton, telephone interview with the author, February 18, 2010.

127 *Look, for instance:* Lile Jia, Edward Hirt, and Samuel Karpen, "Lessons from a Faraway Land: The Effect of Spatial Distance on Creative Cognition," *Journal of Experimental Social Psychology* 45 (2009): 1127–31.

128 *In a 2009 study:* William Maddux and Adam Galinsky, "Cultural Borders and Mental Barriers: The Relationship Between Living Abroad and Creativity," *Journal of Personality and Social Psychology* 96 (2009): 1047–61.

130 *her memoir Dream Doll:* Ruth Handler, *Dream Doll: The Ruth Handler Story* (New York: Longmeadow Press, 1995), 3.

133 *"When you finish":* Zadie Smith, "That Crafty Feeling," *Believer,* June 2008.

134 *Once that happens:* Stanislas Dehaene, *Reading in the Brain: The New Science of How We Read* (New York: Penguin, 2009).

6. The Power of Q

140 *Ben Jones, a professor*: Benjamin Jones, "The Burden of Knowledge and the 'Death of the Renaissance Man': Is Innovation Getting Harder?" *Review of Economic Studies* 76 (2009): 283–317.

Brian Uzzi, a sociologist: Brian Uzzi and Jarrett Spiro, "Collaboration and Creativity: The Small World Problem," *American Journal of Sociology* 111 (2005): 447–504.

145 *The only way*: The best sources on the history of Pixar are David Price, *The Pixar Touch* (New York: Vintage, 2009), and Karen Paik, *To Infinity and Beyond!: The Story of Pixar Animation Studios* (New York: Chronicle Books, 2007).

146 *"It was very"*: Ed Catmull, telephone interview with the author, May 1, 2010.

"We were just": Alvy Ray Smith, telephone interview with the author, April 28, 2010.

149 *"The modern Hollywood"*: Ed Catmull eloquently expanded on this idea in a recent article: Ed Catmull, "How Pixar Fosters Collective Creativity," *Harvard Business Review,* September 2008.

152 *In the early seventies*: Thomas Allen, *Managing the Flow of Technology* (Cambridge: MIT Press, 1977).

153 *A similar lesson*: Serguei Saavedra, Kathleen Hagerty, and Brian Uzzi, "Synchronicity, Instant Messaging and Performance Among Financial Traders," *PNAS* 108 (2011): 5296–301.

156 *Every day at*: Author visits to the Pixar studios, March 30 and 31, 2010.

158 *"Creativity is so delicate"*: Alex Osborn, *Your Creative Power* (New York: Broadway Books, 1948).

"Decades of research": Keith Sawyer, *Group Genius* (New York: Basic Books, 2007), 61.

159 *first empirical test:* David Taylor, "Does Group Participation When Using Brainstorming Facilitate or Inhibit Creative Thinking?" *Administrative Science Quarterly* 3 (1958).

Consider this clever: Charlan Nemeth et al., "The Liberating Role of Conflict in Group Creativity," *European Journal of Social Psychology* 34 (2004): 365–74.

162 *To better understand*: Charlan Nemeth and Margaret Ormiston, "Creative Idea Generation: Harmony Versus Stimulation," *European Journal of Social Psychology* 37 (2007): 524–35.

165 *"Yesterday we saw"*: James Stewart, *DisneyWar* (New York: Simon and Schuster, 2005), 408.

167 *"John and I were"*: Paik, *To Infinity and Beyond!*, 212.

170 *I met Wieden*: Author visit to W+K headquarters, February 15, 2010.

7. Urban Friction

176 *When I met*: Author visit to SoHo studios on June 2, 2009.

179 *Instead of dying out*: *World Urbanization Prospects, 2009 Revision*, United Nations; http://esa.un.org/unpd/wup/Documents/WUP2009_Highlights_Final.pdf.

180 *In an influential*: Robert Lucas, "On the Mechanics of Economic Development," *Journal of Monetary Economics* 22 (1988): 3–42.
     While Lucas didn't: Jane Jacobs, *The Death and Life of Great American Cities* (New York: Vintage, 1992 [1961]).

181 *"Look what"*: Ibid., 4.
     There was Mr. Lacey: Ibid., 66–68.

183 *Look, for instance*: Adam Jaffe, M. Trajtenberg, and Rebecca Henderson, "Geographic Localization of Knowledge Spillovers as Evidenced by Patent Citations," *Quarterly Journal of Economics* 108 (1993): 577–98.

184 *Geoffrey West doesn't*: Author visits to Santa Fe Institute, February 2, 2009, and April 4 to 6, 2009.

186 *After two years*: Luis Bettencourt et al., "Growth, Innovation, Scaling and the Pace of Life in Cities," *PNAS* 104 (2007): 7301–6.

189 *While West and Bettencourt*: Marc Bornstein and Helen Bornstein, "The Pace of Life," *Nature* 259 (1976): 557–59.

194 *AnnaLee Saxenian, a professor*: AnnaLee Saxenian, *Regional Advantage: Culture and Competition in Silicon Valley and Route 128* (Cambridge: Harvard University Press, 1996).

196 *"Every year there"*: Tom Wolfe, "The Tinkerings of Robert Noyce," *Esquire*, December 1983.

197 *"The machines were"*: Steve Wozniak, *iWoz: Computer Geek to Cult Icon* (New York: Norton, 2007). See also http://www.atariarchives.org/deli/homebrew_and_how_the_apple.php.

198 *It's four in*: Yossi Vardi, interview with author, Tel Aviv, May 25, 2010.

199 *In the last decade*: For an excellent summary of the Israeli tech boom, see Dan Senor and Saul Singer, *Start-up Nation: The Story of Israel's Economic Miracle* (New York: Twelve, 2009).

200 *Sergey Brin, the*: Ibid., 202.

202 *Martin Ruef, a sociologist*: Martin Ruef, "Strong Ties, Weak Ties and Islands: Structural and Cultural Predictors of Organizational Innovation," *Industrial and Corporate Change* 11 (2002): 427–49.

206 *"Modern life has"*: Edward Glaeser, telephone interview with the author, April 10, 2010.

207 *As a result:* Elena Rocco, "Trust Breaks Down in Electronic Contexts but Can Be Repaired by Some Initial Face-to-Face Contact," *Proceedings of the SIGCHI Conference on Human Factors in Computing Systems,* 496–502.

A similar lesson: K. Lee, J. Brownstein, R. Mills, and I. Kohane, "Does Collocation Inform the Impact of Collaboration?" *PLoS ONE* 5 (12): e14279.doi:10.1371/journal.pone.0014279.

209 *In contrast, the*: Arie de Geus and Peter Senge, *The Living Company* (Cambridge: Harvard Business School Press, 1997).

8. The Shakespeare Paradox

213 *A few years ago*: http://www.monad.com/sdg/Journal/genius.html.

214 *When William Shakespeare*: My recommended books on Shakespeare and his time: Stephen Greenblatt, *Will in the World: How Shakespeare Became Shakespeare* (New York: Norton, 2004); Marjorie Garber, *Shakespeare After All* (New York: Anchor, 2005); James Shapiro, *A Year in the Life of William Shakespeare: 1599* (New York: Harper Perennial, 2006); and Peter Ackroyd, *Shakespeare: The Biography* (New York: Anchor, 2006).

222 *In 1990, the economist*: Paul Romer, "Endogenous Technological Change," *Journal of Political Economy* 98 (1990): 71–102.

"The thing about": Paul Romer, telephone interview with the author, March 25, 2010.

227 *"The great ages"*: T. S. Eliot, *The Sacred Wood and Major Early Essays* (1920; reprint New York: Dover, 2000), 36.

The New Orleans: Author visit to New Orleans, February 12 through 14, 2011.

230 *Consider a recent*: Erik Westby and V. L. Dawson, "Creativity: Asset or Burden in the Classroom?," *Creativity Research Journal* 8 (1995): 1–10.

231 *While NOCCA encourages*: A. Diamond et al., "Preschool Program Improves Cognitive Control," *Science* 318 (2007): 1387–88.

232 *In recent years*: Angela Duckworth, "Grit: Perseverance and Passion for Long-Term Goals," *Journal of Personality and Social Psychology* 92 (2007): 1087–101.

234 *"People act like"*: Author visit to San Diego, February 22, 2011.

236 *This was demonstrated*: E. B. Bonawitz et al., "The Double-Edged Sword

of Pedagogy: Teaching Limits Children's Spontaneous Exploration and Discovery," *Cognition* 120 (2011): 322–30.

238 *Bill James, the pioneer*: Bill James, "Shakespeare and Verlander," *Slate*, March 30, 2011; http://www.slate.com/id/2289380/.

239 *The numbers are*: J. Hunt and M. Gauthier-Loiselle, "How Much Does Immigration Boost Innovation?," NBER Working Paper no. 14312; http://www.nber.org/papers/w14312.

240 *"What you see"*: Robert Watson, telephone interview with the author, March 18, 2011.

243 *A few years*: P. Azoulay, J. Graff Zivin, and G. Manso, "Incentives and Creativity: Evidence from the Academic Life Science," NBER Working Paper no. 15466; http://www.nber.org/papers/w15466.

246 *Dylan, in* Chronicles: Bob Dylan, *Chronicles* (New York: Simon and Schuster, 2004), 240.

Coda

249 *"I was definitely"*: Teller, interview with the author, Las Vegas, January 8, 2009.

ACKNOWLEDGMENTS

The cover of this book is a fib. You'll find my name there but no one else's, which makes it seem as if these words were written in splendid isolation or that I didn't benefit from the essential input of so many other people. But I did! This book would not exist without their time, generosity, and wisdom.

Let me begin with Amanda Cook. At this point, I'm running out of praise. Amanda has been my editor on all three of my books, and though she's certainly tired of my plodding and swollen drafts, I'm sticking to her like glue. (Preferably one of those really strong 3M glues that hold together golf clubs.) Because Amanda doesn't just fix my broken prose—she helps me think through ideas and stories, helps me figure out what, exactly, I want to say. She's both encouraging and brutally honest, adept at inspiring me to keep on going even as she points out all the wrong places I've gone so far. Every single page of this book has been dramatically improved by her red pen.

I've also benefited from the feedback of many friends and colleagues. Robert Krulwich has always been a hero of mine, but he's also a great reader. After looking at an early draft, Robert wrote me an e-mail that was so full of insight, it kept me busy for six months. Nick Davies at Canongate made many important suggestions, both micro and macro. Paddy Harrington, Charles Yao, Mark Breitenberg, and Jad Abumrad also provided valuable comments. Mike Dudek at Bruce Mau Design did an incredible job creating the graphics. Tracy Roe, the best copyeditor in the world, fixed so many mistakes in the manuscript that, after first seeing her edit, I was rather mortified. (One day, Tracy, I'll learn how to use apostrophes.) Sections of this book benefited from the expert fixes of editors at the various publications I'm lucky to write for. A huge, huge thanks goes out to Mark Robinson and Adam Rogers at *Wired,* Leo Carey and Daniel Zalewski at *The New Yorker,* Mary Turner at *Outside,* and James Ryerson and Alex Star at the *New York Times Magazine.*

Various family members suffered through the slog of my rough prose. Ben Lehrer spent countless hours in a café with me helping me cut unnecessary sentences; Michael Lehrer was the source of much creative inspiration; David Lehrer flagged countless relevant studies; and Rachel Lehrer help me fix some of the hardest narrative problems. My mother, Ariella, has read different incarnations of this book too many times to count. Contra my teenage self, I've learned she's just about always right, so I better listen.

My wonderful agents at the Wylie Agency have made this writing life possible. Sarah Chalfant encouraged me when this book was nothing but a half-baked conjecture, insisting that I keep going. She has suggested stories and connected me with characters—I'm so lucky to have her on my side. Andrew Wylie's support has, at several crucial junctures, been essential. James Pullen has been a joy to work with.

I've also been blessed to share the stories of all the people in this book. My eternal gratitude goes out to everyone who let me ask silly questions, whether it was in the lab or at the movie studio or on a beach in Maui. This book would not exist without their patience and brilliance. I'd like to especially thank all the scientists who conduct the research that allows my job to exist. One of my favorite quotes comes from W. H. Auden, who once said that when he found himself in a roomful of scientists, he felt like a "shabby curate who has strayed by mistake into a drawing-room full of dukes." I live that feeling every day.

And then there's my wife, Sarah. While writing this book, she put up with my negative moods, frequent reporting trips, and miscellaneous obsessions. I'm in awe of her kindness and smarts, her willingness to read bad drafts late at night (when she still has her own work to finish) and offer incisive comments on everything from my improper use of apostrophes to my interpretation of *Highway 61 Revisited.* She's the best best friend ever.

Rosie! You arrived toward the end of the writing process, and that's probably a good thing, since you're the most lovely and adorable distraction. I can't wait to see what you become, although I'm already certain that you're the best thing I'll ever help create.

INDEX

Adams, Anne, 104–6
ADHD. *See* attention deficit hyper-
  activity disorder
Adolphe, Bruce, 84–86
*The Adventures of André and Wally
  B* (cartoon), 146
Agee, James, 54
airplane invention, 39
Akinola, Modupe, 77–78
Akiskal, Hagop, 80
alcohol, 49–50. *See also* mixology
Allen, Tom, 152–53, 154
Allen, Woody, 232
Allen curve, 152–53
alpha wave rhythm, 30–31, 36. *See
  also* relaxation
amphetamines. *See* Benzedrine
Anderson, Darla, 150, 152
Andreasen, Nancy, 78, 79–80
anger, 161n
anterior superior temporal gyrus
  (aSTG), 17–18
anthropologist phase, xii–xiii,
  xvii–xviii
Apatow, Judd, 101

Apollonian creativity, 64–65
Apple Computer, 197–98
archetypes of creativity, 64
Archimedes, 7
architecture, 52n, 182n
  Pixar studio design and, 149–53,
    155–56
Aristotle, 76
art. *See* graphic design
Ascham, Roger, 214n
Asperger, Hans, 96–97
Asperger's syndrome, 95–97
Athens, and clustering of genius, 213,
  214
attention. *See also* daydreaming;
    focus; persistence
  alcohol and, 49–50
  amphetamines and, 53–55,
    58–61
  brain anatomy and, 34–35
  dopamine and, 59–61
  drawing as thinking and, 69–70
  loop of creativity and, 65–68
  pupil dilation and, 61n
  sadness and, 76–77

269

frustration. *See* problem phase
Fry, Arthur, 46–48, 49
funding
  for research, 242–43
  venture capital and, 199
Funk, Joshua, 101–2

gamma-wave rhythm, 17–18
Gennaro, Peter, 144
Gick, Mary, 43–44
Gilmore, Gary, 173
Glaeser, Edward, 206
Glaser, Milton, 68–70, 72–73, 79, 232
  Brooklyn Brewery logo project,
    73–75
  New York ad campaign and, 70–72
Goldman, William, 144
Google, 30, 30n, 37, 39, 152
Graham, Ron, 55
graphic design, 69–72
Greenblatt, Stephen, 226n
Greene, Graham, 54
Greene, Robert, 226
Greenwich Village, 180–82, 190,
  196
grit. *See* persistence
group creativity. *See also* horizontal
  interactions
  breakthroughs and, 164–65
  criticism and, 157–63
  musicals and, 140–44
  as necessity, 139–40
  office interactions and, 152–53,
    170
  Pixar and, 152
  scientific teamwork and, 140
  traders' interactions and, 153–55
  Wieden+Kennedy and, 170–74
Guinea, teenagers in, 223
Gutenberg, Johannes, 39

Handler, Ruth, 130–32
Harold and the Purple Crayon (chil-
  dren's book), 38
Heidegger, Martin, 72

HHMI. *See* Howard Hughes Medical
  Institute
High Tech High School (San Diego),
  234–35
Hindy, Steve, 73, 74n
Holinshed, Raphael, 218
Holyoak, Keith, 43–44
Homebrew Computer Club in Silicon
  Valley, 196–98
hopelessness. *See* problem
  phase
horizontal interactions. *See also*
  group creativity; outsider
  thought; sharing of ideas; so-
  cial networks
  Silicon Valley and, 194–98
  3M culture and, 36–44
Howard Hughes Medical Institute
  (HHMI), 242–43
Hudson Street. *See* Greenwich Vil-
  lage
human "collisions." *See also* horizon-
  tal interactions
  Greenwich Village and, 181–82
  as meta-idea, 239–41
Hume, David, 39
Hyde, Lewis, 245

"(I Can't Get No) Satisfaction" (Rolling Stones song), 108
"I Can't Help Myself" (Four Tops
  song), 20–21
ICQ (online chat program), 200,
  204
ideas. *See also* meta-ideas; sharing of
  ideas
  cities as source of, 176–79,
    183–84, 188
  connections among, 8–13
  as inexhaustible resource, 222
  sources of, 139
  vomiting of, 19–22, 64
IDF. *See* Israel Defense Forces
I ♥ NY ad campaign, 70–72
immigration, 239–41